T0319869

The Neuroscience of Organizational Behavior

The Neuroscience of Organizational Behavior

Constant D. Beugré

Professor of Management, Delaware State University, USA

Edward Elgar
PUBLISHING

Cheltenham, UK • Northampton, MA, USA

Published by
Edward Elgar Publishing Limited
The Lypiatts
15 Lansdown Road
Cheltenham
Glos GL50 2JA
UK

Edward Elgar Publishing, Inc.
William Pratt House
9 Dewey Court
Northampton
Massachusetts 01060
USA

A catalogue record for this book
is available from the British Library

Library of Congress Control Number: 2017959497

This book is available electronically in the Elgaronline
Business subject collection
DOI 10.4337/9781783475544

ISBN 978 1 78347 553 7 (cased)
ISBN 978 1 78347 554 4 (eBook)

Typeset by Columns Design XML Ltd, Reading
Printed and bound in Great Britain by TJ International Ltd, Padstow

Contents

Figures and tables

Figures

Tables

About the author

Constant D. Beugré is a Professor of Management at Delaware State University, College of Business, where he teaches courses in Entrepreneurship, Organizational Behavior and International Management at the undergraduate level and Organizational Leadership and Behavior in the MBA program. Dr. Beugré earned a Ph.D. in Management from Rensselaer Polytechnic Institute and a Ph.D. in Industrial/Organizational Psychology from the University of Paris X Nanterre/France.

His research interests include Organizational Justice, Entrepreneurship, and Organizational Neuroscience. He has published six books and more than 80 refereed journal articles, book chapters, and conference proceedings. His publications have appeared in academic outlets, such as *Organizational Behavior and Human Decision Processes, International Journal of Human Resource Management, International Journal of Manpower, Journal of Applied Behavioral Science, Journal of Applied Social Psychology, Journal of General Management, Research in the Sociology of Organizations*, and the *International Journal of Business Venturing*.

Preface

The role of neuroscience in understanding human behavior in organizations has been receiving increasing attention in recent years. To underscore the importance of this role, several authors have coined the terms organizational cognitive neuroscience, neuro-organizational behavior, or organizational neuroscience to emphasize the use of neuroscientific methods in explaining organizational phenomena. In this book, I explore the role of neuroscience in organizational behavior. The book draws from disciplines such as neuroscience, neuroeconomics, cognitive psychology, social cognitive neuroscience, and organizational behavior. Hence, organizational neuroscience is a multidisciplinary field.

I would like to thank those who have provided support for this endeavor. My gratitude goes to my wife and my children, Chris, Constant Junior, and Jane-Victoria for their love and affection. Many thanks to Karissa Venne, the contact person at Edward Elgar Publishing for her patience and understanding. I also thank colleagues and attendees at conferences who provided valuable feedback on some of the work included in this book. All mistakes and shortcomings are mine.

Constant D. Beugré, PhD
May 31, 2017

Introduction

Currently, there is increasing interest in using neuroscientific tools and techniques to study human behavior in organizations. Perhaps this is the consequence of Lee, Senior, and Butler's (2012a) observation that "there has been a missing cognitive neuroscience level in organization studies" (p. 213). Several authors have used terms such as organizational cognitive neuroscience (Butler and Senior, 2007; Lee and Chamberlain, 2007), neuro-organizational behavior (Beugré, 2010), and organizational neuroscience (Becker and Cropanzano, 2010; Becker, Cropanzano, and Sanfey, 2011) to emphasize the application of neuroscience to organizational behavior. Using neuroscientific tools and techniques to study organizational phenomena is encouraging because understanding how the human brain functions could help to better address some of the questions that current organizational science methodologies cannot resolve. In this regard, organizational neuroscience could complement the organizational sciences. Hence, "researchers should see neuroscience as another tool in the toolbox, one that complements existing methods and is mutually informative and enriching" (Becker and Cropanzano, 2010, p. 1059). However, to benefit from findings in neuroscience, organizational scholars must have a solid grasp of both neuroscience and organizational science.

As a field of scientific inquiry and practical relevance, organizational neuroscience is still in its infancy (Beugré, 2010; Butler et al., 2016; Waldman, Ward, and Becker, 2017). The first articles in the mainstream management and organizational behavior journals that specifically mention the application of neuroscience tools and techniques to organizational behavior were mostly published in the first part of this decade. The articles by Becker and Cropanzano (2010), Becker et al. (2011), Senior, Lee, and Butler (2011), and Lee, Senior, and Butler (2012b) were published in the *Journal of Organizational Behavior* and the *Journal of Management* and *Organization Science*. The only publication prior to 2010 was the journal article by Butler and Senior on organizational cognitive neuroscience in 2007. Despite its emerging nature, however, the future of organizational neuroscience looks promising.

The purpose of this book, therefore, is to lay the groundwork for developing the field of organizational neuroscience. In so doing, the book follows the emerging trend of combining the studies of biology and human behavior in the workplace (Heaphy and Dutton, 2008; Akinola, 2010; Colarelli and Arvey, 2015). In fact, the use of biology to explain how people act or react in the workplace has a long history and can be traced back to the work of Hans Selye (1956) on stress and its physiological consequences. We must also acknowledge that more than 80 years ago, Lashley (1930) pointed out the role of neuroscience in explicating human behavior and discussed the concept of "localization of function" of the brain, implying that individual neurons are specialized for particular functions. He specifically notes that "the final explanation of behavior or of mental processes is to be sought in the physiological activity of the body and, in particular, in the properties of the nervous system" (Lashley, 1930, p. 1).

Since then, neuroscientists have based their studies on this notion to localize particular regions of the brain responsible for specific behaviors. Recently, the study of the physiological dimensions of employee behavior has expanded to neuropsychology and particularly the understanding of the impact of brain structures on work and economic behavior, such as decision making, trust, fairness, leadership, and emotions, to name but a few.

In this book, I take on the challenge of describing how a combination of neuroscience and the organization sciences could lead not only to a new discipline but also, most importantly, advance our understanding of how we as human beings make decisions and interact with one another in organizational settings. The book draws from several fields including neuroscience, organization science, neuroeconomics, cognitive psychology, and social cognitive neuroscience. However, for the promising future of organizational neuroscience to materialize, organizational scholars must familiarize themselves with the theories and models of cognitive psychology and the tools and techniques of neuroscience. This could help rephrase the debate on human behavior in the workplace, thereby leading scholars to ask new questions or reformulate old questions in new ways.

In writing this book, my goal is not to contend that organizational neuroscience is a panacea, nor am I advocating that it is a new field that will revolutionize research and practice in management and organizational behavior. Rather, my goal is to contribute to the nascent field of organizational neuroscience by discussing and summarizing the neural basis of the major topics discussed in organizational science, including fairness, motivation and rewards, trust and cooperation, leadership, decision making, creativity and innovation, morality and ethical behavior, emotions, and unconscious bias. Discussing the neural basis of these topics could provide insights to our understanding of the different facets of organizational life. As Becker et al. (2011, p. 951) suggest, "the perspectives from organizational neuroscience can help scholars resolve conceptual disagreements. Issues that are difficult to differentiate at one level of analysis may become more distinctive at the level of neural processing."

Likewise, Senior, Lee, and Butler (2011) add that neuroscience can help realize the relationship between "organizational behavior and our brains and allows us to dissect specific social processes at the neuro-biological level and apply a wider range of analysis to specific organizational research" (p. 804). Volk and Köhler (2012) note that by uncovering neural functioning as the basis of an observable behavior, neuroscience techniques can (1) help researchers form better theories about the underlying reasons for observable behaviors in a given context; and (2) create more accurate tests of these theories than if they were to use other, less direct means, such as standard self-report measures.

Organizational neuroscience provides an opportunity for cross-pollination of organizational research and opportunities for targeted interventions and could also help us better understand the role of training and retraining in neuroplasticity (Damasio, Everitt, and Bishop, 1996; Bechara, Damasio, and Damasio, 2000). Neuroplasticity refers to the ability of the brain to rewire itself based on the environment in which the individual operates. Organizational neuroscience should not limit itself to brain mapping. Rather, it must draw sound conclusions from studies using neuroscience tools and techniques. It is only by doing so that it can make new and useful contributions to existing theories in the organizational sciences. To some extent, "human behavior at some level is biological, but this is not to say that biological reductionism yields a simple, singular, or satisfactory explanation for complex behaviors or that molecular forms of representation provide the only or best level of analysis for understanding human behavior" (Cacioppo, 2002, p. 820).

The book is divided into ten chapters. The first chapter presents and discusses the nature of organizational neuroscience. The second chapter describes the methods used in neuroscientific research that could be useful for organizational neuroscience scholars. The third chapter analyzes the neural basis of decision making. The fourth chapter explores the neural basis of creativity and innovation in organizations. The fifth chapter analyzes the neural foundations of motivation and rewards. The sixth chapter deals with issues of leadership and their neural foundations. The seventh chapter addresses the neural foundation of fairness. The eighth chapter describes how neuroscience could help us understand trust and cooperation in organizations. The ninth chapter discusses the neural basis of ethics and morality. Finally, the tenth chapter addresses the neural basis of emotions and unconscious bias in organizations. Together, these different topics represent those that are extensively studied in research in organizational behavior and neuroscience and that are largely discussed in organizational behavior textbooks.

1. The nature of organizational neuroscience

Organizations can be construed as social cognitive systems populated by people who gather, compute, analyze, and interpret information, and interact with one another. These social cognitive systems cannot be studied without reference to the neural substrates that underlie cognitions. Thus, this chapter presents the nascent field of organizational neuroscience to the reader. In doing so, it provides a working definition of organizational neuroscience, discusses its interdisciplinary nature, explores its levels of analysis, and identifies some of the research questions it could help elucidate.

1 UNDERSTANDING ORGANIZATIONAL NEUROSCIENCE

In recent years, there has been increasing interest in applying neuroscientific methods and techniques to the study of organizational phenomena (Butler and Senior, 2007; Lee and Chamberlain, 2007; Senior, Lee, and Butler, 2011; Becker and Cropanzano, 2010; Beugré, 2010; Becker, Cropanzano, and Sanfey, 2011; Butler, 2014). Beugré (2010) introduced the construct of neuro-organizational behavior, which he defined as the study of the impact of the brain on behavior that occurs in organizations (p. 289). Likewise, Butler and Senior (2007), Lee and Chamberlain (2007), and Senior et al. (2011) introduced the field of organizational cognitive neuroscience to explain the role of neuroscience in human behavior in organizations. Lee and Chamberlain (2007, p. 22) defined organizational cognitive neuroscience as "the study of the processes within the brain that underlie or influence human decisions, behaviors, and interactions either a) within organizations or b) in response to organizational manifestations or institutions". This definition is similar to the one provided by Butler and Senior (2007) who conceived of organizational cognitive neuroscience as the use of "neuroscientific methods to analyze and understand human behavior within the applied setting of organizations. This application may be at the individual, group, organizational, and interorganizational levels" (p. 8).

Lee, Senior, and Butler (2012b) make a distinction between organizational neuroscience (ON), social cognitive neuroscience (SCN), and organizational cognitive neuroscience (OCN). Specifically, they contend that organizational neuroscience focuses on brain anatomy and structures, whereas SCN and OCN deal with multiple levels of analysis and are interested in the interplay between biological systems and cognitions. The authors acknowledge an overlap between organizational neuroscience and SCN and OCN. Particularly, they argue that "OCN is symbiotic with ON as well as SCN, to form a detailed theoretical framework that helps scholars to understand the complexities of the social behavior that occurs within organizations" (p. 923).

In this book, I consider that the distinction between OCN and organizational neuroscience is a superficial one because organizational neuroscience does not limit itself to the description of neural anatomy. Rather, it encompasses both a description of brain structures as well as an understanding of the neural mechanisms that substantiate cognitions. My argument is that both organizational neuroscience and organizational cognitive neuroscience study the same phenomena and use the same research tools. For example, the neural basis of topics such as decision making, emotions, cognitions, trust, cooperation, leadership, and ethics are studied by both disciplines using the same neuroscientific methods.

Becker and Cropanzano (2010) coined the term organizational neuroscience, which they conceive "as a deliberate and judicious approach to spanning the divide between neuroscience and organizational science" (p. 1055). They suggest that "existing theories of organizational behavior can be enhanced by incorporating the findings and themes from neuroscience regarding how the brain produces cognition, attitudes, and behaviors" (ibid.). Becker et al. (2011, p. 937) view organizational neuroscience as an "interpretive framework that sheds new light on existing problems as well as highlighting problems that otherwise might not have been considered." Thus, organizational neuroscience could help to answer questions that current organization science research cannot.

The three terms, organizational cognitive neuroscience, neuro-organizational behavior, and organizational neuroscience, have been used to describe the same field. Although this may seem awkward, it is hardly surprising because organizational neuroscience does not even exist as a unified field. Thus, the aim of this book is to contribute to the development of a new field that blends together neuroscience and organizational science. To this end, it is important to consider a unified appellation of the field – common sense and human practice dictate that a new baby must be named. Organizational scholars cannot create a field combining neuroscience and the organization sciences if they do not know what this new field is called or cannot delineate it. As a consequence, I make a modest effort to provide a first step in this direction.

In doing so, I favor the term organizational neuroscience used by Becker and Cropanzano (2010) and Becker et al. (2011) to encompass what I described as neuro-organizational behavior (Beugré, 2010) and what Butler and Senior (2007) called organizational cognitive neuroscience or social cognitive neuroscience. I favor this term for two main reasons. First, the term organizational neuroscience tends to be broader than organizational cognitive neuroscience, where the latter could be misleading. Indeed, organizational cognitive neuroscience may appear at first glance to favor conscious efforts made by individuals and overlook social aspects, emotions, or unconscious phenomena. The term organizational neuroscience is also broader than the term neuro-organizational behavior because neuro-organizational behavior could be limited to human behavior in existing organizations, and overlook the application of neuroscience to fields such as entrepreneurship, management information systems, and strategic management.

Second, the term organizational neuroscience falls under the same categories as the terms neuroeconomics and neuromarketing. These latter fields combine the tools of neuroscience with existing theories and models of economics or marketing. Thus, the field of organizational neuroscience aims to combine tools and techniques of neuroscience with organizational science theories.

To make organizational neuroscience a coherent field, organizational scholars must provide a clear definition of the field and identify pertinent research questions that can only be explored by this nascent field. Indeed, "a scientific concept has meaning only because scientists mean something by it. The meaning is scientifically valid only if what they intend by it becomes actual: problems are solved and intentions are fulfilled as inquiry continues" (Kaplan, 1964, p. 46). I therefore define organizational neuroscience as the *field that integrates the use of neuroscientific methods and techniques to the study of organizational phenomena*. Defined as such, organizational neuroscience is a multidisciplinary field.

2 ORGANIZATIONAL NEUROSCIENCE AS A MULTIDISCIPLINARY FIELD

Organizational neuroscience can be construed as a multidisciplinary field that draws from disciplines such as neuroscience, neuroeconomics, social cognitive neuroscience, cognitive psychology, and neuroscience. The multidisciplinary nature of organizational neuroscience was advocated by Beugré (2010) who introduced a neuro-organizational behavior paradigm, which he described as a multidisciplinary discipline that draws its knowledge and methods from cognitive psychology, neuroeconomics, neuroscience, organizational behavior, and social cognitive neuroscience. In the following paragraphs, I explore the extent to which each of these five disciplines contributes to the nascent field of organizational neuroscience. Other disciplines not identified here could also benefit organizational neuroscience but I focus on the five illustrated in Figure 1.1.

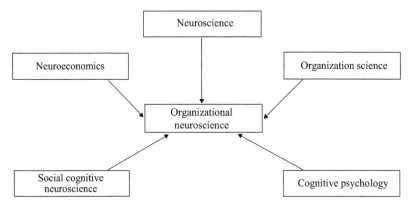

Figure 1.1   Disciplines contributing to organizational neuroscience

2.1   Neuroscience

Neuroscience is the study of the brain, its structure, functions, and how it affects human behavior. Neuroscience also helps our understanding of how to repair brain structures and restore brain functions when they are impaired. Neuroscience itself is an interdisciplinary approach stemming from biology, chemistry, mathematics, medicine, physics, and psychology. As Mukamel and Fried (2012, p. 511) put it, "the ultimate goal of neuroscience research is to understand the operating mechanism of the human brain and to exploit this understanding to devise methods for repair when it malfunctions."

   Kable (2011) identified three ways in which neuroscience can inform our understanding of human cognition and behavior: (1) the use of association tests; (2) the use of tests of necessity; and (3) the use of tests of sufficiency. First, the use of tests of association involves observing or experimentally manipulating psychological states or behavior, simultaneously measuring neural activity, and examining correlations between the two. Second, using a test of necessity involves disrupting neural activity and showing that this manipulation impairs a specific behavior or psychological function. Third, the use of tests of sufficiency involves enhancing neural activity and showing that this manipulation results in a specific behavioral or psychological state. Kable's conceptualization could encompass the use of neuroscientific tools such as functional magnetic resonance imaging (fMRI), electroencephalography (EEG), magnetoencephalography (MEG) and transcranial magnetic stimulation (TMS) to investigate particular organizational phenomena. Neuroscience can help develop "basic processing models that can be used to generate predictions about individual and group performance" (Paulus et al., 2009, p. 1085).

"Neuroscience holds great promise for advancing organizational theories and practice" (Cropanzano and Becker, 2013, p. 310), particularly as "neuroscientific methods provide the most reliable window into the unconscious brain that is currently available" (ibid.). Although neuroscience can provide useful knowledge for organizational scholars, the potential for the misuse of its methods is real. We are already witnessing the overselling of the promise of neuroscientific methods in marketing and management (Cropanzano and Becker, 2013; Ashkanasy, Becker, and Waldman, 2014).

2.2 Organization Science

Organizational science is loosely defined as the set of disciplines that study the functioning of human organizations and their well-being. It is an interdisciplinary field, including human resources management, industrial and organizational psychology, organizational behavior, organizational theory, strategic management, and management. It also draws from other social science disciplines such as psychology, sociology, political science, economics, and anthropology. Thus, any study of how human organizations function can fall under the purview of organizational science.

Of particular importance to organizational neuroscience is the field of organizational behavior, which refers to the study of human behavior in organizations. Organizational science does not face the same problem as economics insofar as it does not assume that humans are rational. In fact, it has always given a prominent role to subjective aspects of human behavior and decision making such as emotions and moods. Organizational scholars and psychologists have long contended that behavior in organizations results from the interplay between the individual and the environment (Lewin, 1947). Therefore, the behavior displayed cannot be analyzed independently of the context in which it occurred. Issues related to decision making, social perceptions, moods and emotions, organizational change, creativity and innovation, culture, personality, attitudes, and behaviors are the focus of research in organization science.

As illustrated in Figure 1.1, the disciplines contributing to organizational neuroscience overlap but also focus on particular issues that are not always studied by organizational scholars. For example, neuroscientists are not concerned with the application of their findings to organizations. The same is true for neuroeconomists, social cognitive neuroscientists, and cognitive psychologists who are not necessarily concerned with the application of their findings to the workplace. However, their findings could be used by organizational scholars to understand organizational phenomena and provide guidelines for practitioners. A recent discipline that has taken advantage of the increased development in neuroscience is neuroeconomics (Camerer, 1999; Camerer and Loewenstein, 2004). Although still an emerging discipline, neuroeconomics could provide valuable insights for organizational neuroscience scholars to help them navigate the difficulties that they may face as they transform organizational neuroscience into a legitimate discipline in the organizational sciences.

2.3 Neuroeconomics

Neuroeconomics refers to the study of the impact of the brain on economic decisions (Glimcher, 2003; Camerer, Loewenstein, and Prelec, 2004; Zak, 2004; Innocenti and Sirigu, 2012). It is interdisciplinary in that it combines neuroscientific measurement techniques and economic methods and theory to investigate how the human brain generates decisions in economic and social contexts (Zak, 2004; Braeutigam, 2005; Fehr, Fischbacher, and Kosfeld, 2005). After all, the human mind is the driver of all economic action (Braeutigam, 2005) and any study of human economic behavior cannot ignore the brain.

One of the goals of neuroeconomics is to explore the unobservable, subjective aspects of decision making in economic situations (Camerer, Bhatt, and Hsu, 2007), thereby overlapping with fields such as psychology and organizational behavior. Some of the pioneers of neuroeconomics, such as Camerer et al. (2004), contend that "in a strict sense, all economic activity must involve the human brain" (p. 555). Neuroeconomics is a subfield of behavioral economics and experimental economics – it addresses the neural foundations of behaviors that behavioral economists often study; it also involves elaborate experiments using neuroscientific tools to study economic phenomena. For instance, issues of revealed preferences, time discounting, and valuation, to name but a few, are studied using neuroscience methods.

Neuroeconomics also uses game theory (Von Neuman and Morgenstern, 1944; Camerer, 2003) to study phenomena such as trust, cooperation, fairness, and others. The most popular games used in neuroeconomics research are the Ultimatum Game (Güth, Schmittberger, and Schwarze, 1982; Thaler, 1988), the Prisoner's Dilemma Game (Fehr and Camerer, 2007; Sanfey, 2007), and the Trust Game (Kreps, 1990; Berg, Dickhaut, and McCabe, 1995; Johnson and Mislin, 2011).

The Ultimatum Game includes two players: a proposer and a responder. The first player, the proposer, receives a sum of money, generally $10 dollars in most experiments, and decides how to split it between him- or herself and the other player, generally known as the responder. The responder may decide to accept or reject the offer. If the offer is accepted, the money is split according to the proposal. However, if the offer is rejected, both players end up empty-handed. This game is used in experimental economics, behavioral economics, and now neuroeconomics to study issues of fairness. Very often, responders reject offers less than 30 percent of the amount.

In the Prisoner's Dilemma Game, two prisoners are arrested for a crime allegedly committed. The two prisoners are in separate cells and do not have any means to communicate between them. Because the prosecutor lacks sufficient evidence for a conviction of the two prisoners, he or she approaches them individually and offers a bargain. The prosecutor gives each prisoner the opportunity either to betray the other by testifying that the other committed the crime, or to cooperate with the other by remaining silent. Each prisoner is faced with three choices: (1) if prisoner A and prisoner B, each betray the other, each of them serves two years in prison; (2) if prisoner A betrays prisoner B but prisoner B remains silent, prisoner A will be set free and prisoner B will serve three years in prison (and vice versa); (3) if prisoner A and prisoner B both remain silent, both of them will only serve one year in prison (on the lesser charge). Very often, each prisoner decides to betray the other, when in fact each would be better off by cooperating. The matrix displayed in Table 1.1 illustrates the possible outcomes of the Prisoner's Dilemma Game. There are several versions of the game, which is often used in behavioral economics and neuroeconomics to study cooperation and reciprocity.

Table 1.1 Payoffs in the Prisoner's Dilemma Game

	Prisoner B Cooperates	Prisoner B Betrays
Prisoner A Cooperates	Each prisoner serves one year	Prisoner A gets three years' sentence Prisoner B goes free
Prisoner A Betrays	Prisoner A goes free Prisoner B gets three years' sentence	Each prisoner gets two years' sentence

Berg et al. (1995) developed the Trust Game, which is played as follows. Subjects in Room A decide how much of their $10 show-up fee to send to an anonymous counterpart in Room B. Subjects are informed that each dollar sent would triple by the time it reaches Room B. Subjects in Room B then decide how much of their tripled money to keep and how much to send back to their respective counterparts. The unique Nash equilibrium prediction for this game with perfect information is to send zero money. However, the authors found that this prediction was rejected because 30 out of 32 Room A subjects sent money. This contradiction could be explained by the fact that trust is an important component of exchange relationships. The trust game is used to study trust, cooperation, reputation, and reciprocity.

Sanfey (2007) argues that by combining the models and tasks of game theory with modern psychological and neuroscientific methods, the neuroeconomic approach to the study of social decision making has the potential to extend our knowledge of brain mechanisms involved in social decisions and to advance theoretical models of how we make decisions in a rich, interactive environment. Lee (2008) emphasized the use of game theory in neuroeconomics by arguing that decision making in a social group has two distinguishing features. First, humans and other animals routinely alter their behavior in response to changes in their physical and social environment. As a result, the outcomes of decisions that depend on the behavior of multiple decision makers are difficult to predict and require highly adaptive decision making strategies. Second, decision makers may have preferences regarding consequences for other individuals and therefore choose their actions to improve or reduce the well-being of others.

According to Camerer (2007), the largest innovation may come from pointing to biological variables that have great influence on behavior and are underweighted or ignored in standard economic theory. The neuroeconomic theory of the individual replaces the (perennially useful) fiction of a utility-maximizing individual who has a single goal with a more detailed account of how components of the individual – brain regions, cognitive control, and neural circuits – interact and communicate to determine individual behavior (p. C28).

Neuroeconomics has primarily challenged the standard economic assumption that decision making is a unitary process, a simple matter of integrated and coherent utility maximization, suggesting instead that it is driven by the interaction between automatic and controlled processes (Loewenstein, Rick, and Cohen, 2008). Volk and Köhler (2012) argue that neuroeconomics can help researchers "form better theories about the underlying reasons for observable behaviors in a given context and create more accurate tests of their theories than if they were to use other, less direct measures such as standard self-report measures" (p. 523). By doing so, neuroeconomics can contribute to theory pruning (Leavitt, Mitchell, and Petterson, 2010).

The same reasoning could apply to organizational neuroscience. Indeed, neuroscience could help prune theories in organizational sciences. Neuro-economics studies phenomena that are relevant for organizational neuro-science. For example, neuroeconomists study topics such as decision making, uncertainty, risk, trust, and cooperation, even the emotional basis of human behavior. Findings from neuroeconomists could find applications in organizations. Indeed, organizations are entities populated by people who make decisions on a regular basis. In addition, organizational members cooperate with one another to accomplish certain tasks. Hence, there should be a close collaboration between neuroeconomists and organizational neuroscience scholars.

2.4 Social Cognitive Neuroscience

Social cognitive neuroscience is closely related to social neuroscience, and is defined as "an interdisciplinary field devoted to understanding how biological systems implement social processes and behavior, capitalizing on biological concepts and methods to inform and refine theories of social processes and behavior, and using social and behavioral concepts and data to refine theories of neural organization and function" (Cacioppo et al., 2007, p. 100). According to Lieberman (2007a), social cognitive neuroscience examines social phenomena and processes using cognitive neuroscience research tools such as neuroimaging. Knowledge from social cognitive neuroscience helps to accomplish the following: (1) understanding others; (2) understanding oneself; (3) controlling oneself; and (4) the processes that occur at the interface of self and others.

However, the field of social neuroscience could not ignore the role of social cognitions in social and interpersonal relations. Thus, the field of social cognitive neuroscience developed in large part as a recognition of the role of social cognitions in social neuroscience. Social cognition refers to "the ability to construct representations of the relations between oneself and others, and to use those representations flexibly to guide social behavior" (Adolphs, 2001, p. 231). Hence, it encompasses any cognitive process that involves conspecifics, either at a group level or on a one-to-one basis (Blakemore, 2004, p. 216). This is particularly relevant to organizations because they represent social entities where people interact with one another on a regular basis.

Social cognitive neuroscience is a multidisciplinary field embedded in the social sciences because it draws from social cognition, political cognition, behavioral economics, and anthropology (Lieberman, 2007a). It allows people to understand themselves and others (Lieberman, 2007a). Lieberman focuses on social cognitive neuroscience as a means of studying social behavior using the tools of cognitive neuroscience, a field that combines cognitive psychology and neuroscience. Social cognitive neuroscience combines the "tools of cognitive neuroscience with questions and theories from various social sciences" (Lieberman, 2007b, p. 260).

According to Bechtel (2002, p. S49), the dominant perspective in cognitive neuroscience revolves around two principles: (1) different brain areas perform different information processing operations and (2) an explanation of a cognitive performance involves both decomposing an overall task into component information processing activities and determining what brain area performs each. Ochsner and Lieberman (2001) consider social cognitive neuroscience as "an emerging interdisciplinary field of research that seeks to understand phenomena in terms of interactions between 3 levels of analysis: the social level, which is concerned with the motivational and social factors that influence behavior and experience; the cognitive level, which is concerned with the information-processing mechanisms that give rise to social-level phenomena; and the neural level, which is concerned with the brain mechanisms that instantiate cognitive-level processes" (p. 717, abstract).

Ochsner and Lieberman (2001) represented these three levels of analysis by a prism (Figure 1.2) intended to capture the idea that social cognitive neuroscience is about studying phenomena at many levels of analysis to learn how and when brain systems are used to mediate motivated human behavior (Ochsner and Lieberman, 2001, p. 719). The task of cognitive neuroscience is to map the information-processing structure of the human mind, and to discover how this computational organization is implemented in the physical organization of the brain. It

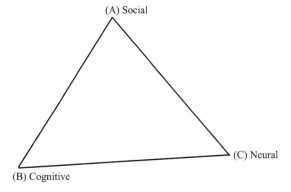

Figure 1.2   The prism of social cognitive neuroscience

helps address questions such as, how do brain events give rise to mental and behavioral events? Because organizations are populated by people and involve their interactions, findings from social cognitive neuroscience could enrich organizational neuroscience by providing knowledge related to the neural basis of social and interpersonal relations.

Because social cognitive neuroscience uses neuroscience and cognitive psychology to study social behavior, it has implications for social behavior in organizational settings. Subsequently, Butler and Senior (2007) and Lee and Chamberlain (2007) have expanded it to organizational contexts. For proponents of a social cognitive neuroscience perspective of organizational behavior, human behaviors in organizations can better be understood if one makes sense of their neural foundations. Such an approach offers some interesting insights because knowledge of the brain can provide useful information about how people react toward others and understand the organizational world they navigate on a regular basis. For example, organizations can be perceived as arenas of emotion-production; that is, people act and react to others and events based on their emotional appraisal.

We know that emotions play a critical role in how people make decisions and act in organizations. This is hardly new, but social organizational neuroscience can provide further refinements. Thus, organizational neuroscience draws from social cognitive neuroscience insofar as it acknowledges the role of the neural basis of cognitions in organizational behavior. It can build on previous research on social cognitive neuroscience and apply it to organizational life. Organizational neuroscience builds on the scientific developments in organizational behavior and the recognition of the role of cognition and emotion in explicating human behavior at work.

2.5　Cognitive Psychology

Cognitive psychology is the study of how people mentally represent and process information. As such, it includes within its domain mental abilities such as perception, learning, memory, reasoning, problem solving, and decision making (Sternberg, 1981, p. 1181). The focus on cognitions had led to the development of theories on the role of cognition on decision making under risk and uncertainty. It has also led to the development of new theories such as prospect theory (Kahneman and Tversky, 1979) and cognitive biases in decision making (Tversky and Kahneman, 1974).

Cognitive psychology could play an important role in organizational neuroscience for at least three reasons. First, the cognitive revolution in psychology has allowed a focus on cognitions and thoughts (Sperry, 1993; Miller, 2003). In fact, how people react to situations depends on how they construe them. In this approach, mental abilities are considered key factors in influencing human behavior. Cognitive psychology contributed to the development of artificial intelligence (Newell and Simon, 1976; Simon, 1980). Computers could replicate human thoughts and make decisions. One of the limitations of previous theories of cognitive psychology was the role of emotions in human actions. Because cognitions could not be integrated smoothly with emotions, scholars introduced the construct of embodied cognition, which focuses on the importance of action and on how action shapes perception, the self, and language (Glenberg, Witt, and Metcalfe, 2013).

Second, the development of the construct of socially situated cognition (Neisser, 1967; Sperry, 1993; Mandler, 2002; Miller, 2003; Smith and Semin, 2004) indicates that organizational phenomena could be studied through the lens of cognition, behavior, and context. Smith and Semin (2004, p. 53) highlight four assumptions that are common to socially situated cognition: (1) cognition is for the adaptive regulation of action, and mental representations are action oriented; (2) cognition is embodied, drawing on our sensorimotor abilities and environments as well as our brains; (3) cognition and action are the emergent outcome of dynamic processes of interaction between an agent and an environment; and (4) cognition is distributed across brains and the environment.

Third, most of the topics studied in organizational science involve the processing of information: "All human mental events occur as the result of neural information processing" (Ilardi and Feldman, 2001, p. 1072). This statement may imply that the organizational neuroscience paradigm suggests that human behaviors in organizations occur as the result of neural information processing. For example, people constantly make decisions that are choices based on how they evaluate the situation and the resources available. People also decide who to trust or not. These examples demonstrate the potential contribution that cognitive psychology could make to organizational neuroscience.

Cognitive psychology helps us to understand the mental structure of thoughts and human behavior. Poldrack (2006) argued that "the goal of cognitive psychology is to understand the underlying mental architecture that supports cognitive functions" (p. 59). As a result, most research in cognitive psychology involves the manipulation of tasks to determine which cognitive functions are affected. Cognitive psychologists also explore the types of cognitive functions and abilities that can help individuals effectively perform certain tasks. As Poldrack notes, "if neuroimaging were able to provide information regarding what cognitive processes were engaged in performance of a particular task, cognitive psychologists would have gained a powerful new tool" (ibid.). Cognitive psychology is also concerned with the role of automatic and controlled processing in cognition and its impact on human behavior (Sternberg, 1981; Lieberman, 2007a, 2007b).

3 LEVELS OF ANALYSIS OF ORGANIZATIONAL NEUROSCIENCE

3.1 Individual Level of Analysis

If organizational neuroscience is to become a legitimate field of study in the organizational sciences, we must identify the levels of analysis it could address. Should it focus only on the individual level of analysis or should it also address group and/or organizational levels of analysis? (See Figure 1.3, which illustrates all three.) Beugré (2010) suggests that neuro-organizational behavior includes three levels of analysis: neural, cognitive, and behavioral. The neural level focuses on identifying the brain regions that are activated when organizational members display particular types of behavior. The cognitive level involves internal mental processes that rely on these neural substrates such as memory and information processing (Lee and Chamberlain, 2007). The behavioral level concerns observable actions displayed by organizational members.

The neural level deals with how different brain structures function and interact to influence human attitudes and behaviors. Questions related to the role of specific brain structures could be relevant here. This is important because one needs to clearly understand the specific role of particular brain structures before identifying which ones influence the behavior of interest. At the cognitive level, researchers are interested in knowing how brain structures affect the thought processes we experience. Such an understanding could provide clues for knowing how people process information and make decisions. The behavioral level is the observable part and can help us understand how we act or react in particular ways when faced with some environmental stimuli. All three levels focus on the behavior of individuals acting independently. However, we know that in organizations, people tend to work in groups and teams to perform specific tasks. Can organizational neuroscience address the functioning of such units of analysis?

3.2 Group and Social Interaction Level of Analysis

At this level, organizational scholars might look at the impact of neuroscience on group functioning and social interactions within organizations. Understanding of neural science concepts and methods could help explain how interpersonal dynamics occur in organizations. For example, issues of hidden biases and diversity could be studied to understand whether they are rooted in brain structures. Likewise, topics such as social cognition, shared cognitions, shared mental models, and intergroup relationships, including intergroup conflicts, could be studied through the lens of neuroscience. The discovery of the mirror neuron system, a group of specialized neurons that mirrors the actions and behavior of others (Rizzolatti and Craighero, 2004; Rizzolatti and Fabbri-Destro, 2008), could help shed light on the neural basis of social interactions in organizations.

3.3 Organizational Level of Analysis

Research in organizational behavior focusing on the level of the entire organization has included topics such as organizational politics, organizational structure and design, organizational culture, and change. An organizational-level analysis of organizational neuroscience could entail the study of the neural basis of the topics identified above. For example, researchers could study the neural basis of organizational culture, change, or political behavior in organizations. In fact, understanding the neural underpinnings of behavior can help explain group- or organizational-level phenomena because "many phenomena in organizations have their theoretical foundation in the cognition, affect, behavior, and characteristics of individuals, which, through social interaction, exchange, and amplification, have emergent properties that manifest at higher levels. In other words, many collective constructs represent the aggregate influence of individuals" (Kozlowski and Klein, 2000, p. 15).

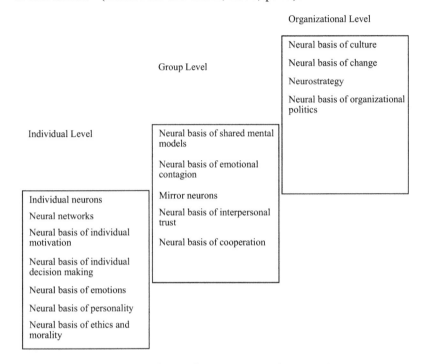

Figure 1.3   Levels of analysis of organizational neuroscience

4   PROSPECTS FOR ORGANIZATIONAL NEUROSCIENCE

What questions can organizational neuroscience address? What topics can better be studied as a result of using neuroscientific tools and techniques? How can an understanding of neuroscience lead to better management practices? This section addresses these questions.

4.1   Contribution to Research

Organizational neuroscience can contribute to theory development and the analysis of organizational topics in a new light. For example, topics currently studied in the organizational sciences such as leadership, motivation, decision making, trust, creativity, and change could be expanded with neuroscientific tools. The use of these tools could help better understand these organizational phenomena.

Becker and Cropanzano (2010) suggested three areas where organizational neuroscience could provide useful insights to organizational scholars: (1) combatting procrastination; (2) mirror neurons and group subclimates; and (3) attitudes and behaviors. To explain the role of organizational neuroscience in combatting procrastination, the authors use the dual-process theory (implicit, automatic processing and explicit, controlled processing). Humans are vulnerable to procrastination because of the existence of an automatic system that tends to retain them in the status quo and away from new targets and behaviors. The prefrontal cortex plays an important role in the control system that allows for the planning and deliberate focus on activities.

The mirror neuron system (MNS) presents evidence that people unconsciously react to the actions of others and could play a central role in social interactions and teamwork in organizations. According to Rizzolatti and Craighero (2004) and Rizzolatti and Fabbri-Destro (2008), the mirror neuron system responds to visual observation and facial expressions. People tend to mimic actions performed by others. Hence, this system could help to advance social learning and explain vicarious learning. People learn by observing others. Becker and Cropanzano (2010, p. 1057) contend that "when team members interact, the MNS is finely tuned to perceive the actions, expressions, and body language of others. They implicitly learn from and assess the behaviors of other members. The MNS will lead group members who interact frequently to converge toward attitudes and behaviors that are adaptive for the group, but not necessarily for the organization." In other words, the mirror neuron system could explain how good or bad behavior spreads within teams and even organizations.

Becker and Cropanzano (2010) also mentioned the role of neuroscience in explaining how implicit attitudes can influence explicit attitudes and behaviors in organizations. Using the dual-process theory, they contend that "an implicit attitude is rapid, automatic, and comprises unconscious evaluative response to stimuli, whereas an explicit attitude is a relatively slower, deliberative conscious evaluation based on contextual information" (p. 1058).

Other authors have also emphasized the potential benefits of organizational neuroscience (Beugré, 2010; Powell, 2011; Ward and Becker, 2013). Ward and Becker (2013) identified four benefits of organizational neuroscience: (1) it offers several tools for data collection; (2) it offers alternatives to self-report questionnaires; (3) it offers opportunities for strengthening construct validation; and (4) provides new ways to refine existing theories.

4.2 Contribution to Management Practice

An understanding of organizational neuroscience could provide guidelines for management practice in several areas. For example, Ward, Volk, and Becker (2015) identified several applications of organizational neuroscience, including training and development, job design, high-performance assessment, motivating communication, and conflict prevention. It is worth mentioning that several domains could be added to those identified by Ward and collaborators. For instance, one could use findings from organizational neuroscience to further creativity, team building, emotional control and regulation, the development of trust, decision making, ethical behavior, and diversity in the workplace. In fact, the remaining chapters of this book, after discussing the research tools in the next chapter, deal with the role of neuroscience in understanding these topics and providing guidelines for management practice.

Rangel, Camerer, and Montague (2008) note that organizational neuroscience can be useful to artificial intelligence: "A question of particular interest is which features of the brain's decision-making mechanisms are optimal and should be imitated by artificial systems, and which mechanisms can be improved on" (p. 555). In an era of big data analytics and algorithms, findings from organizational neuroscience could provide useful guidelines for managers and organizations. It could also be useful for self-development. Indeed, individuals could use findings from brain science to improve their lives as individuals and economic agents. There is evidence that exposure to brain functioning and especially one's own brain images tends to improve self-regulation: "Neuroscience technology, combined with brain education and training, can improve self-control at its source in the brain, and improve experiences, expressions and effectiveness" (Powell and Puccinelli, 2012, p. 210).

Despite its potential benefits, organizational neuroscience still faces several challenges. These challenges are often related to the tools used by researchers. The following chapter discusses neuroscientific tools that are relevant for research in organizational neuroscience. It will also address the challenges organizational neuroscience faces as a nascent discipline.

2. Methods of organizational neuroscience

For organizational neuroscience to yield its promise, it must rely on sound research methods that should offer opportunities for reliability, validity, and generalizability. These methods should also complement those currently used in the organizational sciences. Researchers in the nascent field of organizational neuroscience use neuroscientific methods to study organizational phenomena. These "methods test the association between brain activity and behavior, the necessity of brain activity for behavior, and the sufficiency of brain activity for behavior" (Kable, 2011, p. 64). Although organizational neuroscience uses the tools and techniques of neuroscience, not all of them are relevant for studying organizational phenomena.

This chapter presents a brief overview of five major neuroscientific tools that could be used or are currently used in organizational neuroscience research. They include functional magnetic resonance imaging (fMRI), electroencephalography (EEG), magnetoencephalography (MEG), positron emission tomography (PET), and transcranial magnetic stimulation (TMS). Today, these five neuroscientific tools represent the most widely used functional neuroimaging modalities. The chapter also addresses the challenges faced by the nascent field of organizational neuroscience. In fact, the prospects or limitations of a scientific paradigm depend on the methods it uses to study the topics of interest. Sound and reliable research tools can lead to the "discovery" and understanding of the key paradigms explored by the field. However, poor research methods may prevent better analysis of the main research questions.

1 FUNCTIONAL MAGNETIC RESONANCE IMAGING

Functional magnetic resonance imaging (fMRI) is the most popular method used to measure brain activity and make inferences about human behavior. For example, Kable (2011) reports that 60 to 70 percent of all empirical studies that have applied neuroscience to the study of decision making have used just one technique, fMRI. According to Kable (2011), there are two reasons that underlie the frequent use of fMRI in neuro-scientific studies. First, scholars use fMRI as a research technique because of its superior spatial resolution. Second, the widespread availability of fMRI equipment in research institutions including MRI scanners and user-friendly software for the analysis of fMRI data makes it a prominent and easy choice for neuroscientific research.

Another argument that could facilitate the use of fMRI is the persuasive nature of brain images themselves. By showing vivid pictures of the brain, neuroscientific methods such as fMRI tend to persuade the reader, even the most educated ones. Colorful and scientific-looking brain images also serve a purpose in their own right within organizations as a powerful tool of persuasion (Spence, 2016). McCabe and Castel (2008) conducted three experiments that showed that presenting brain images with articles summarizing cognitive neuroscience research resulted in higher ratings of scientific reasoning for arguments made in those articles, as compared to articles accompanied by bar graphs, a topographical map of brain activation, or no image. These data lend support to the notion that part of the fascination, and the credibility, of brain imaging research lies in the persuasive power of the actual brain images themselves. Such images can provide credibility that the science behind the picture is genuine. As O'Connor, Rees, and Joffe (2012, p. 220) put it, "logically irrelevant neuroscience information imbues an argument with authoritative, scientific credibility."

Studies that use fMRI provide direct observation of specific brain regions that are activated during the experiments. In addition, "functional magnetic resonance imaging provides the best combination of spatiotemporal resolution and anatomical coverage" (Kable, 2011, p. 67). The predominant variable associated with the use of fMRI is the blood oxygenation level-dependent (BOLD) signal, which is commonly used to assess changes in blood flow in the brain in reaction to various stimuli (Waldman, Ward, and Becker, 2017). Consequently, fMRI remains an important tool for research on the neural basis of organizational phenomena.

Despite its popularity, fMRI presents some limitations. Dulleck et al. (2011) note that "the sheer magnitude of the technology precludes simultaneous observation of a large number of participants and imposes limitations on the experimental design" (p. 118). One must also acknowledge that fMRI tools limit the opportunity for field research, because one has to conduct the study while the subject is lying in a scanner. Thus, the environment in which fMRI studies are conducted could be considered an artificial one. Despite these limitations, fMRI remains one of the most popular methods used by researchers. For example, increasing attention to extracting signals from specified regions of interest (ROIs) is moving fMRI further forward.

Because of these limitations, "the results of fMRI procedures should always be used in conjunction with other technologies to guard against issues such as reverse inference" (Butler et al., 2016, p. 544). (I will discuss the issue of reverse inference in a later section.) FMRI "should only be considered an indirect measure of neuronal activity because the time it takes for dynamic changes to occur in blood flow is much longer than that for neurons to fire off their electrochemical messages. As such, fMRI is said to have poor temporal resolution" (Balthazard and Thatcher, 2015, pp. 94–95).

2   ELECTROENCEPHALOGRAPHY

Electroencephalography (EEG) is one of the oldest tools used to measure brain activity (Handy, 2005; Luck, 2005; Kable, 2011). Hans Berger (1929) first described the EEG signal generated by alpha (8–12 Hz) activity when he demonstrated that closing the eyes decreased sensory input and increased alpha power over the occipital scalp. Berger considered EEG as an opportunity to look inside the "black box," the brain. EEG reflects brain cells' electrical activity. In conducting an EEG study, an active electrode on the scalp measures the electric field that is generated by the sum of the momentary post-synaptic potentials of the brain (Michel and Murray, 2012).

In EEG, brain waves are sampled and analyzed to detect neural activity. Long used by psychologists, it consists of placing electrodes on the scalp to measure brain activity. Studies that use EEG focus on skin conductance (SC) measurements, feedback negativity (FN) (also called medial frontal negativity, MFN), and event response potential (ERP). EEG can allow the use of larger samples than fMRI. However, EEG is not as spatially precise as fMRI but its temporal precision is superior. For these advantages, EEG is becoming an important tool for research in organizational neuroscience (Waldman et al., 2011b; Balthazard et al., 2012). Although EEG is not invasive, it does not measure brain activity directly.

In addition to "traditional" EEG, researchers often use qEEG (quantitative electroencephalography) in which they compare expected values based on normative databases of neurologically normal individuals to that of the subject or subjects of a particular study. It is worth mentioning that the spatial resolution of qEEG is less than that of fMRI, although some organizational neuroscience scholars (Hannah et al., 2013; Waldman,

Wang, and Fenters, 2016) consider its spatial resolution as adequate in organizational phenomena. According to Balthazard and Thatcher (2015), qEEG has excellent temporal resolution, adequate spatial resolution, is relatively inexpensive, and its use involves no health risk (p. 97).

Lorensen and Dickson (2003, p. 55) note that "the development of quantitative EEG analysis has afforded researchers the opportunity to scientifically investigate whether large samples of participants who meet medical and psychological criteria for normality also display stable, reliable and common patterns in their EEG." There are several qEEG normative databases and selecting the ones that are appropriate for a specific study could be a daunting task for researchers. The databases must be updated to reflect developments in research. EEG can be used in conjunction with other methods. For example, Mayhew et al. (2013) used EEG and fMRI and found that spontaneous EEG alpha oscillation interacts with positive and negative BOLD responses in the visual-auditory cortices and default-mode network.

3 MAGNETOENCEPHALOGRAPHY

Cohen (1968) presented evidence of the existence of a smaller fluctuating magnetic field around the head produced by alpha-rhythm currents often found in EEG. Whereas EEG electrodes record the scalp potential, magnetoencephalography (MEG) sensors measure the magnetic field outside the head. The magnetic fields are produced by electrical activity in the brain. Cohen (1972) compared the brain magnetic fields of four normal participants and one abnormal participant and found that the normal MEG showed the alpha rhythm, as did the EEG, when participants' eyes were closed. However, this MEG also showed higher detector sensitivity, by a factor of 3, which would be necessary to clearly show the smaller brain events when the eyes were open. The abnormal MEG, including a measurement of the direct current component, suggests that the MEG may yield some information that is new and different from that provided by the EEG (Cohen, 1972).

EEG consists of recording the brain's electric fields (Berger, 1929), whereas MEG (Cohen, 1972) records the brain's magnetic fields. MEG allows the simultaneous recording of the brain's magnetic activity from large arrays of sensors covering the whole head. Unlike EEG, which records the electrical fields of the brain, MEG records magnetic fields from outside the brain. "A change in an electrical field induces a magnetic field and a change in a magnetic field induces an electrical field" (Stam, 2010, p. 128). The EEG and the MEG are very close methodologies since the main sources of both kinds of signals are essentially the same, ionic currents generated by biochemical processes at the cellular level" (Da Silva, 2013, p. 1112). MEG shares with EEG the advantages of directly measuring brain activity and having a very high temporal resolution, which is only limited by the sample frequency of the electronics. Both are also non-invasive and can help increase the sample size in research in organizational neuroscience.

Converting the currents in the coils to a signal that can be measured requires another delicate piece of technology: superconducting quantum interference devices or SQUIDs. These consist of "small rings with a very small interruption, which can be crossed by electrons thanks to quantum effects" (Stam, 2010, p. 129). SQUIDs are the key elements in MEG systems. "Compared to EEG, MEG has the following advantages: (i) in contrast to the electrical field, the magnetic field is hardly affected by intervening tissues such as the skull; (ii) measurements of the magnetic field do not require a reference as is the case with EEG; this is an advantage, in particular in studies of functional connectivity; (iii) measurements from a very large number of sensors are relatively more easy with MEG than with EEG, since the tedious application of electrodes on the scalp is not necessary" (ibid.).

In MEG, the SQUIDs are coupled to the brain magnetic fields using combinations of superconducting coils called flux transformers (primary sensors) (Vrba and Robinson, 2001, p. 249). MEG signals are measured on the surface of the head and reflect the current flow in the functioning brain (Vrba and Robinson, 2001, p. 250). The SQUID sensor is the heart of the MEG system. The rationale for using MEG to study brain activity and cognitive processes is that "the physics of magnetic measurement permit three-dimensional localization of current sources and the changes in ionic source currents can be studied on a time scale of less than 1ms [milliseconds]. MEG can be used for functional neuroimaging of events that are not accessible either to functional MRI or to nuclear imaging methods" (Vrba and Robinson, 2001, p. 269). MEG has both good spatial and excellent temporal resolution (Cornelissen et al., 2009).

MEG, like EEG, provides an excellent, millisecond-scale time resolution, and allows the estimation of the spatial distribution of the underlying activity, in favorable cases with a localization accuracy of a few millimeters (Ahlfors and Mody, 2016). MEG has an excellent temporal resolution. Fornasier and Pitolli (2008, p. 386) note that "the aim of magnetoencephalography is the analysis of brain functionality through the measurements of the tiny magnetic fields generated by neuronal currents." MEG is a non-invasive technique and its goal is to accurately determine the current density flowing within the volume of the head in the working human brain. In brain studies, "the critical issue is not simply to localize cognitive functions to some site in the brain but to find out the patterns of dynamic interaction between different brain systems underlying a cognitive process" (Da Silva, 2013, p. 1123). Braeutigam (2014) contends that MEG can be a powerful research tool in organizational neuroscience. Specifically, he argues that magnetoencephalography "can allow the detailed mapping of brain activity associated with complex cognitive processes" (p. 744).

4 POSITRON EMISSION TOMOGRAPHY

A positron emission tomography (PET) study begins with the injection or inhalation of a radiopharmaceutical. In a PET study, participants are injected with a radioactive tracer [$H_2^{15}O$] while they are performing a cognitive task (Cabeza and Nyberg, 1997, p. 1). The PET scanner detects the distribution of the tracer in the brain, which is an indication of the pattern of blood flow. A PET scan generally lasts between 40 and 70 seconds. Because of this, the research should be repeated after several minutes to yield reliable results. "A standard PET experiment consists of 6–10 scans per subject, made 10–15 minutes apart" (ibid.).

Methods such as EEG, MEG, and PET do not provide the researcher with the opportunity to directly observe brain regions. Rather, EEG provides information about electrical activity in the brain through which the researcher could determine which brain regions could be implicated. The same is true for studies using PET. In such studies, researchers alter the level of a neurotransmitter, serotonin, to assess participants' reactions. The level of serotonin in the brain helps the researcher to infer the brain regions that might have influenced participants' responses.

"Positron emission tomography (PET) scanning provides the ideal tool for measuring the changes in brain activity that occur when a task becomes automatic, because scans can be performed at different times during learning" (Jenkins et al., 1994, p. 3775). "The field view of the PET scanner makes it possible to examine functionally related areas that may have a wide anatomical distribution" (ibid.). The tracer C50, can be used to map the changes in regional cerebral blood flow (rCBF) that accompany changes in neuronal activity.

One of the advantages of PET is its flexibility. As Lammertsma (1992, p. 113) observed, "positron emission tomography (PET) allows for accurate non-invasive in vivo measurements of regional tissue function. This ability rests on (1) the decay characteristics of positron emitters; and (2) the range of tracers which can be labelled with a positron-emitting radionuclide. In PET, positron emitters decay by the emission of a positron, an electron with positive charge."

Since a positron is unstable, it will travel at most a few millimeters in tissue before combining with an electron. This annihilation process results in the emission of two gamma rays, each with an energy of 511 kilo-electron volts, travelling in opposite directions. These two gamma rays can be detected simultaneously by two opposing coincidence detectors, resulting not only in the registration of the event but also of the line along which this event has taken place (line of response). Studies involving PET as a methodology have used cerebral blood flow as a measure of brain activity. The major problems with PET are its cost, the short half-life of most positron-emitting isotopes, which means an on-site cyclotron is required, and the scanners themselves are significantly more expensive than single-photon cameras (Ollinger and Fessler, 1997, p. 44).

5 TRANSCRANIAL MAGNETIC STIMULATION

Transcranial magnetic stimulation (TMS) is a non-invasive technique that is used to create a sort of virtual lesion (Lee and Chamberlain, 2007) to assess the effects of these virtual lesions on the behaviors studied (Lee and Chamberlain, 2007; Kable, 2011). It is traditionally used to interfere with cognitive processes by disrupting the functioning of the stimulated area. Barker, Jalinous, and Freeston (1985) introduced the use of TMS in the study of neural activity and cognitive neuroscience. TMS follows in the footsteps of lesion studies used by psychologists to study the inability of brain patients to perform certain tasks. In their review of the neuroscientific methods used in organizational cognitive neuroscience, Lee and Chamberlain (2007) identified the advantages and shortcomings of TMS. They considered that the lesions created for the purpose of a given study are transient, which implies that the brain cannot develop compensation strategies.

The virtual lesion can also be tightly controlled and researchers can control the "before/after lesion" effect, which "provides the opportunity to eliminate between-subject variation and error" (Lee and Chamberlain, 2007, p. 28). TMS can be a mapping tool for studying perceptual, motor, and cognitive functions in the human brain as well as provide the opportunity for investigating the causal implication of an area in a specific task (Veniero et al., 2016). Despite these advantages, the use of TMS alone is not likely to offer significant insight into many of the key organizational research areas (Lee and Chamberlain, 2007, p. 28).

TMS modulates brain activity, which in turn modulates cognitive behavior. In TMS, a stimulating coil is positioned over the participant's head to deliver a strong and transient magnetic pulse to induce an electric current at the cortical surface. "TMS can be delivered as a single pulse (spTMS) at a precise time point or as a series of stimuli in conventional or patterned protocols" (Veniero et al., 2016, p. 2). TMS is part of what are called non-invasive brain stimulation (NIBS) methods. TMS intends to modify brain activity and hence influence cognitive processes. It is relatively low cost, easy to apply, and could help conduct causal studies rather than correlational ones, which are often conducted when researchers use fMRI or EEG.

The effects of TMS could be weak and short-lived. In TMS studies, "the stimulating coil is held over a subject's head and as a brief pulse of current is passed through it, a magnetic field is generated that passes through the subject's scalp and skull with negligible attenuation (only decaying by the square of the distance)" (Pascual-Leone, Walsh, and Rothwell, 2000, p. 232). It is also possible to apply a series of pulses at rates of up to 50 Hz (this is known as repetitive TMS, or rTMS).

Another version of TMS is the transcranial direct current stimulation (tDCS). This procedure is non-invasive and involves the application of a low-intensity constant direct current that enters the brain via the anode and leaves the tissue via the cathode. "During tDCS, a weak electric current (1mA [milliampere]) is applied using two surface electrodes" (Poreisz et al., 2007, p. 208). tDCS is an established brain stimulation method allowing the induction of excitability enhancements or reductions of a targeted brain area in a polarity-dependent manner (Horvath, Forte, and Carter, 2015; Veniero et al., 2016). tDCS of different cortical areas results in modifications of perceptual, cognitive, and behavioral functions. As such, tDCS can be a valuable research tool in organizational neuroscience. TMS, tDCS, and rTMS may have some minor adverse effects, such as headache, nausea, and an itching sensation, which must be acknowledged and addressed. For example, Poreisz et al. (2007) found that tDCS applied to motor and non-motor areas according to the present tDCS guidelines is associated with relatively minor adverse effects.

Several brain methods can be used in a single study. Walsh and Conwey (2000, p. 77) note that the combination of TMS with PET and fMRI has been useful in studying the connectivity of the human brain, in validating the specificity of TMS, and in guiding the location of TMS application. Komssi and Kähkönen (2006) combined the use of TMS and EEG and observed that neuronal activation is induced with remarkably low stimulation intensities. The authors also showed that TMS–EEG enables the study of interhemispheric connections with high spatio-temporal specificity and the assessment of cortical reactivity with excellent sensitivity. The key feature of TMS is its unique ability to painlessly induce neuronal activation in the intact brain.

Table 2.1 summarizes the neuroscientific tools discussed above, but this list of methods is far from exhaustive. As technology develops, other neuroscientific methods may be relevant for research in organizational neuroscience. Waldman et al. (2016) note that neuroscience methods, especially those associated with EEG technology, are becoming more user friendly, affordable, and practical for organizational research.

It is also important for organizational scholars to familiarize themselves with the use of these tools. This would require retraining organizational scholars willing to undertake research in organizational neuroscience but also collaboration with colleagues from neuroscience departments or colleagues doing research in social science disciplines such as neuroeconomics and social cognitive neuroscience who use neuroscientific tools. Indeed, the potential of neuroscience as a viable framework for studying human behavior in organizations depends on scholars' ability to evaluate, design, analyze, and accurately interpret neuroscientific research (Jack et al., 2017).

Table 2.1   Neuroscientific tools useful for research in organizational neuroscience

Tools	Key Features	Strengths	Weaknesses
Functional magnetic resonance imaging (fMRI)	Detects magnetic field distortions that can result from changes in cortical activity, chemical composition of the brain or tissue damage. Imaging of brain function tends to use changes in oxygenated blood (the BOLD contrast)	Non-invasive, easy to use, able to locate the source of the signal very accurately (excellent spatial resolution)	Expensive, latest technology not available to all, poor temporal resolution as signal changes occur, post-cortical activity, subjects exposed to loud noise due to rapid switching on/off of radiofrequency signal, can be stressful

Tools	Key Features	Strengths	Weaknesses
Electro-encephalography (EEG)	Detects (at scalp surface) electrical potential differences derived from neural activity. Subjects have a number of electrodes attached to their heads	Widely available, non-invasive, long history of use leading to well-accepted body of theory, excellent temporal resolution	Does not measure neural activity directly, signal must travel through tissue and skull to surface, relies on the accurate modeling of this path, unable to conclusively determine the actual local of activation in the brain leading to poor spatial resolution
Magneto-encephalography (MEG)	Detects magnetic signals associated with the electrical activity in the brain. Subjects sit in large scanner	Non-invasive, able to directly measure signals of neural activity, thus more robust than EEG, new techniques can minimize the "inverse problem," excellent temporal resolution	Expensive to purchase, use, and maintain equipment, suffers from the inverse problem, standardized equipment size can make it unsuitable for some subjects
Positron electron tomography (PET)	Detects the distribution of the tracer in the brain. Begins with the injection or inhalation of a radiopharmaceutical	Non-invasive and flexible Measures the changes in brain activity that occur when a task becomes automatic Examines functionally related areas that may have a wide anatomical distribution	Does not provide the researcher with the opportunity to directly observe brain regions Relatively expensive Short half-life of most positron-emitting isotopes
TMS (rTMS, tDCS)	Used to create a sort of virtual lesion A magnetic field is generated that passes through the subject's scalp and skull with negligible attenuation	Non-invasive Could allow the study of causal implications Can be used as a mapping tool for studying perceptual, motor, and cognitive functions in the human brain	May not offer significant insight into many of the key organizational research areas Minor side-effects

Source: Partially adapted from Lee and Chamberlain (2007, p. 27).

6   CHALLENGES OF ORGANIZATIONAL NEUROSCIENCE

As a nascent field, organizational neuroscience faces several challenges. One of the prominent critics of organizational neuroscience is Dirk Lindebaum (2013a, 2013b, 2016), who argues that neuroscience as applied in management can have negative consequences. Specifically, he argues that studies that suggest "remedial measures" for employees and managers who do not conform to a "so-called neurological profile" are incomplete (Lindebaum, 2013b). Reasoning in terms of remedial measures would imply that healthy participants, such as employees and managers, present some "neurological deficiencies" that must be addressed. Doing so poses two main issues: organizational neuroscience scholars are not equipped with the expertise and knowledge to make such a determination and such a conclusion raises important ethical issues.

Lindebaum (2016) also argues that organizational neuroscience relies on a "supply and demand" perspective of management theory development. He specifically notes that accelerating demands for novel theories in management studies means that new methodologies and data are sometimes accepted prematurely. This may lead to a rush to publish studies using neuroscientific methods. Lindebaum (2016) suggests that the interest for research in neuroscience in management stems from the direction of public research funding (research on neuroscience is likely to be funded) and the likelihood to publish neuroscientific research in management journals to boost their impact factor. Looking at the supply side, he notes three issues: (1) the statistical power of studies using fMRI (the "gold" standard of research in organizational neuroscience) is unacceptably low; (2) the use of imprecise "motherhood" statements (that is, statements that lack clear conceptual and theoretical directions enabling a better understanding of how precisely brain networks account for, and are influenced by, behaviors that have practical relevance in the context of work; p. 542); and (3) the dismissal of ethical concerns in the formulation of management theories and practice informed by neuroscience.

In this section, I have organized the concerns raised by Lindebaum and other scholars into five main challenges that need to be addressed if organizational neuroscience can gain legitimacy as a field of scientific inquiry. They include: (1) reverse reference; (2) issues of reductionism; (3) sample size and statistical power; (4) organizational neuroscience as a management fad; and (5) ethical issues.

6.1 Reverse Inference

The issue of reverse inference is inherent in the tools and techniques of neuroscience and not specific to organizational neuroscience. In reverse inference, the engagement of a particular cognitive process is inferred from the activation of a particular brain region (Poldrack, 2006, p. 59). For example, using fMRI to study the neural basis of human behavior necessarily raises issues of reverse inference. As Lee, Senior, and Butler (2012b, p. 922) put it, "a fundamental limitation in neuroimaging is an inability to infer complex social behavior from observations of specific activated brain regions." In reverse inference, "if a region is activated by a large number of cognitive processes, then activation in that region provides relatively weak evidence of the engagement of the cognitive process; conversely, if the region is activated relatively selectively by the specific process of interest, then one can infer with substantial confidence that the process is engaged given activation in the region" (Poldrack, 2006, p. 60). The problem with reverse inference is that it "reflects the logical fallacy of affirming the consequent" (Poldrack, 2008, p. 223).

Lee et al. (2012b, p. 926) introduced the "reverse inference paradox," which suggests that the more complex a cognitive task, the less chance there is of identifying a discrete cortical response. Lee et al. (ibid.) argue that "it is impossible to be confident in the assignment of a particular cognitive task A to a particular active region of the brain X, outside of the possibility that other processes also may have contributed to activation of region X during task A." In the reverse inference model, the activation is used as evidence that region X is associated with the task (ibid.).

Recently, Jack et al. (2017) note that reverse inference in organizational neuroscience research may not be as bad as anticipated. They proposed four design elements that enable solid reverse inference: (1) hypothesis-driven research; (2) careful and simple designs; (3) taking a broader view of the brain and cognition; and (4) the use of tailored designs such as parametric modulation studies, performance-related analyses, cognitive conjunction, studies of the neural effects of behavioral intervention, and prospective prediction studies. In addition to these design elements, the authors suggest that inferences from the brain to cognition/behavior can be improved using neural intervention methods, such as transcranial magnetic stimulation and lesion studies. Jack et al. (2017) also note that because there is not a strict one-to-one mapping between brain areas and cognitive processes, the use of reverse inference without appropriate checks and controls limits the ability of researchers to draw valid conclusions regarding the relationship between a specific task and the putatively engaged psychological process on the one hand and a specific brain area on the other.

Lee et al. (2012b) also introduced the concept of forward inference, which "refers to an analytic approach that rather than simply looking for activation in a particular region of the brain, uses patterns of brain activation to distinguish between competing cognitive theories" (Lee et al., 2012, p. 926). "Forward inference refers to the use of qualitatively different patterns of activity over the brain to distinguish between competing cognitive theories" (Henson, 2006, p. 64). In forward inference, the only assumption is that the same cognitive process is not supported by different brain conditions within the same experiment (ibid.). Lee et al. (2012b) suggest that combining forward and reverse inference paradigms has great potential to advance management science (p. 927). This is particularly important because "forward inference and reverse inference are intertwined" (Klein, 2010, p. 188).

In addition to issues of reverse inference and forward inference, some authors (Becker, Cropanzano, and Sanfey, 2011; Ashkanasy et al., 2014) have argued that the use of neuroscience in organizational science can lead to reductionism.

6.2 Issues of Reductionism

Another limitation of organizational neuroscience is the issue of reductionism raised by several authors (Becker et al., 2011). Reductionism means that studies in organizational neuroscience would simply reduce human behavior in organizations to neural mechanisms. Proponents of this criticism argue that organizational behavior is complex and cannot be reduced to neural substrates. For example, Balleine (2007) notes that decision making involves an interaction between the cognitive, motivational and behavioral processes and therefore cannot be limited to a single neuron or cell. Ashkanasy et al. (2014) explore the extent to which research in organizational neuroscience tends to espouse a form of reductionism and suggest the use of the term, "networks" to avoid focusing on a simple localization of human behavior. Marshall (2009) argues that forward inference does not raise questions about reductionism, whereas reverse inference does; this is because in forward inference, hypotheses that were previously used are tested using a neuroscientific approach. Reverse inference raises issues of reductionism because it involves an "attempt to localize psychological functions" (Marshall, 2009, p. 116).

However, other scholars contend that organizational neuroscience does not lead to reductionism. Indeed, an organizational neuroscience perspective views the use of neuroscience as combining both the individual brain and the organizational environment in which the individual moves, acts, and reacts. Hence, it is far from being reductionist because it embraces the now accepted notion that cognitions are socially embedded (Healey and Hodgkinson, 2014). Thus, the effect of brain structures on organizational behavior cannot be studied independently of the environment in which the individual operates. From an "embodied perspective, neuroscience is not a reductionist force, but rather a way to relate internal and external aspects of representations through the sensorimotor interface of an organism that is deeply embedded in the world" (Marshall, 2009, p. 122). Therefore, the challenge for organizational researchers, as with neuroscience researchers in general, is to "begin to think more in terms of networks, rather than the strict localization of mental and behavior processes" (Ashkanasy et al., 2014, p. 910).

The work of Hannah et al. (2013) on neural networks in leadership and Healey and Hodgkinson (2014) on socially situated cognition is illustrative in this regard. Healey and Hodgkinson (2014) contend that "if MOS [management and organization studies] are to benefit meaningfully from neuroscience, it must establish a viable means of engaging with theoretical and empirical developments in this rapidly expanding field, without losing sight of the *socially embedded nature* of organizational life" (p. 766, original emphasis). They consider two developments – critical realism (with its emphasis on the fact that reality is socially constructed) and socially situated cognition (cognition is embodied and draws from sensorimotor abilities and environments as well as brains; Smith and Semin, 2004) – as conceptual frameworks that can allow for a better contribution of neuroscience to management and organizational science. Therefore, Healey and Hodgkinson (2014, p. 782) argue that "explaining organizational behavior by ignoring neurophysiological materiality is to divorce organizations from the lower-level boundary conditions that shape their actuality."

Other authors have also contended that a limitation inherent in organizational neuroscience is the overinterpretation of cerebral activation data (Butler et al., 2016). It is worth noting that proponents of organizational neuroscience do not intend to reduce organizational phenomena to neural substrates. As is commonly admitted, in the social sciences, human behavior is complex and represents the result of the interplay between the individual and the environment. Thus, behaviors that are displayed in social settings, such as within organizations, cannot be reduced to some neurons firing while others remain dormant. Reductionism could present serious challenges to the nascent field of organizational neuroscience.

Bechtel (2002) reviewed three techniques commonly used in research in cognitive neuroscience – lesion studies, cell recording, and neuroimaging – and concluded that they can each provide some suggestive clues. However, he advises that each of these techniques must be complemented by the results of other techniques. "Even together, they do not provide a definitive account of the processing in a particular area" Bechtel (2002, p. S57). Based on Bechtel's conclusion, it is probably safe to suggest that triangulation is important in research in organizational neuroscience. "Using several techniques can help address the deficiencies inherent in each technique. Hence, an interdisciplinary organizational neuroscience perspective provides the most promising means of conceptualizing and investigating previously invisible drivers of organizational behavior" (Ashkanasy et al., 2014, p. 913).

6.3   Sample Size Effects and Statistical Power

Research in the organizational sciences often uses large samples for statistical analysis. Very often, large samples offer the advantage of generalizability and reliability of the research instruments used, particularly in the case of survey research. Large samples also facilitate the use of more complex statistical methods, such as multiple regression analysis and structural equation modeling. However, research in organizational neuroscience, specifically, studies using fMRI, PET or TMS, do not offer the opportunity to use large samples. Such research has a limited number of participants per study, which could limit the generalizability of the research findings. Small sample size undermines the reliability of neuroscience findings (Button et al., 2013). Button et al. (2013) also showed that the average statistical power of studies in the neurosciences is very low. The authors also note that a study with low statistical power has a reduced chance of detecting a true effect.

Lindebaum and Jordan (2014, p. 899) identified low statistical power in existing neuroscientific studies, leading to questions over generalizability of results and a fundamental inability to localize phenomena in the brain. In addition, research in organizational neuroscience is conducted in laboratory and controlled settings. Such settings limit the number of participants because of space availability and time constraints. Organizational life is complex and people tend to make decisions in the heat of the moment. Such an opportunity is not always offered to participants of experimental studies in organizational neuroscience. Possibly, using tools such as EEG or MEG may help increase sample sizes; however, these tools may not allow researchers to directly observe brain activity. A solution might be to combine the use of fMRI and EEG or MEG to address the sample size effect.

6.4 Organizational Neuroscience as a Management Fad

Lindebaum and Jordan (2014) raised two main questions when addressing the challenges facing organizational neuroscience: (1) How strong is the science behind organizational neuroscience? (2) Why are we doing this research? The authors contend that localization studies may be misleading because several brain regions may be involved in the occurrence of specific mental states (reverse inference, discussed earlier). It is also important for organizational neuroscience scholars to explain clearly the reasons why neuroscientific methods and techniques should be applied to the study of organizational behavior.

In addition, research in organizational neuroscience may be perceived as a management fad rather than a serious academic endeavor. This criticism is because some trade journals and consultants are overselling the promise of organizational neuroscience findings. Terms such as neuroleadership, neuromanagement, and neuromarketing are becoming overused by consultants providing various services – Legrenzi and Umiltà (2011) coined the word "neuromania" to emphasize the exaggerated nature of using the prefix "neuro" in several disciplines. And, as indicated in the discussion of fMRI as a research tool in organizational neuroscience, brain images tend to be persuasive to readers.

To avoid organizational neuroscience being perceived as a management fad, scholars must ensure that it is rooted in solid scientific evidence and the guidelines prescribed for practice are grounded in sound theories – the use of neuroscientific methods should not be a panacea. This requires the use of both neuroscientific methods as well as theories of organizational science. The potential of neuroscience as a viable framework for studying human behavior in organizations depends on scholars' ability to evaluate, design, analyze, and accurately interpret neuroscientific research (Jack et al., 2017).

Organizational neuroscience researchers could combine several methods. They could combine EEG and fMRI to better identify brain regions of interest when participants perform certain tasks. For example, Mayhew et al. (2013) combined EEG recording with BOLD responses using fMRI. Nevertheless, we must acknowledge that the combination of methods depends not only on the problem studied but also on the availability of resources, and especially financial resources. Because some of the methods used in neuroscience research are relatively expensive, there is a possibility for organizational neuroscience researchers to think in terms of return on investment. Acquiring equipment for research purposes could be considered an investment and the return on investment could be measured not only by the quality of the research but also by its relevance and impact.

6.5   Ethical Implications of Organizational Neuroscience

Organizational neuroscience research has several ethical implications. First, people may oversell the methods and findings of organizational neuroscience, as discussed above. As Cropanzano and Becker (2013) put it, "the most pressing ethical danger is not the methods themselves but rather in people overestimating or overselling the power of these methods" (p. 307). Second, the procedures themselves could raise ethical issues if they are invasive. Third, Lindebaum (2013b) noted that "it is questionable whether participants have freedom to decline participating in research programs" (Lindebaum, 2013b, p. 301). Fourth, organizational neuroscience may lead to management intervention – interventions that may wrongly assume that employees and managers are deficient and therefore need "brain training" or "brain re-adjustments" to become more effective. Such an orientation led Lindebaum to suggest that "there is a possibility that neuroscience will invade the privacy of the human and there could be a neurological modification of the employee" (Lindebaum, 2013b, p. 298). Fifth, research in organizational neuroscience could also raise issues of privacy and confidentiality (Marcus, 2002; Roskies, 2002; Waldman et al., 2016). This concerns how the data should not only be collected but also stored. More importantly, in the case that organizational neuroscience research leads to the existence of neurological diseases, should such information be communicated to participants and their employers (Waldman et al., 2016)? This question is particularly intriguing because there are currently no guidelines on how to deal with such issues.

According to Roskies (2002), the ethics of neuroscience can be roughly subdivided into two groups: (1) ethical issues and considerations that should be raised in the course of designing and executing neuroscientific studies; and (2) the evaluation of the ethical and social impact that the results of those studies might have, or ought to have. The first concerns internal issues related to the ethics of the research. For example, how should research participants be treated? How should results be reported and stored? The second concerns the consequences of studying the neural foundations of human behavior on the participants themselves and on society at large. For example, studying the neural foundations of behavior raises issues of privacy and confidentiality.

It also raises issues of neuro-enhancement; that is, whether research from neuroscience could lead to the development of smart drugs. In such a case, to what extent is it ethical for some people to have access to these drugs to enhance their cognitive capabilities while others do not? It is important to distinguish neuroethics as the study of the ethical implications of neuroscience from the study of the neural foundations of ethics. While the former focuses on whether doing research in neuroscience raises ethical issues, the latter deals with the neural substrates of ethical behavior – whether the sense of right or wrong can be substantiated by brain structures. Indeed, neuroethics is concerned with ethical, legal, and social policy implications of neuroscience, and with aspects of neuroscience research itself (Illes and Bird, 2006).

3. The neural basis of decision making

Decision making is the essence of the manager's job (Mintzberg, 1979) and represents "a deliberate process that results in the commitment to a categorical proposition" (Gold and Shadlen, 2007, p. 535). The most well-known and used decision making model in economics is the rational choice model, which assumes that human beings have infinite knowledge and they can process all information available and act in such a manner as to maximize their expected utility (Von Neumann and Morgenstern, 1944). The rational choice paradigm assumes that decision makers are rational agents who select the option with the best expected utility. However, it ignores the role of emotions and feelings. It also ignores the limited capacity that we as humans have when making decisions.

Simon (1947, 1955, 1956) introduced the concept of bounded rationality to explain the extent to which the decision maker's ability to make rational decisions is limited. The bounded rationality model contends that humans have a limited capacity to process information; therefore, they make decisions that are good enough – they "satisfice" in the terminology of Herbert Simon. For example, emotions, heuristics, and unconscious bias often influence decisions related to hiring, performance appraisal, or promotion. Miller (1994) notes that the span of absolute judgment and the span of immediate memory impose severe limitations on the amount of information that people are able to receive, process, and remember. He suggests that to make effective decisions despite these limitations, people must organize the stimulus input simultaneously into several dimensions and successively into a sequence or chunks, to break (or at least stretch) the information bottleneck.

The emergence of behavioral economics (Allison, 1983; Camerer, 1999; Camerer and Loewenstein, 2004; Sent, 2004) and neuroeconomics (Glimcher, 2003; Camerer, Loewenstein, and Prelec, 2004) is shedding light on the neural foundations of decision making. Decision making is a cognitive process and cognitive control stems from the active maintenance of patterns of activity in the prefrontal cortex that represent goals and the means to achieve them (Miller and Cohen, 2001). According to Miller and Cohen, these patterns of activity provide signals to other brain structures whose net effect is to guide the flow of activity along neural pathways that establish the proper mappings between inputs, internal states, and outputs needed to perform a given task. A neural perspective of decision making assumes that humans are far from being rational. Many decisions are made in the context of social interactions (Rilling, King-Casas, and Sanfey, 2008; Rilling and Sanfey, 2011). This is particularly true for employees and managers who constantly interact with others in organizations. Sanfey et al. (2006) note that despite substantial advances, "the question of how we make decisions and judgments continues to pose important challenges for scientific research" (p. 108).

If there is a single topic on which most research in neuroeconomics has focused, it is decision making. Classical economics and, notably, the rational choice model, has considered decision making as a rational process where the decision maker tries to maximize personal utility. However, research in psychology, behavioral economics, or neuro-economics has rightfully found that humans are not always rational. It is also worth mentioning that the decision making process is integrative because it involves cognitive processes, emotional processes, and the values attributed to choices and their consequences.

1 NEURAL STUDIES OF DECISION MAKING

Neuroeconomics could be described as the discipline in economics and the social sciences that devotes itself to the study of the neural foundations of decision making. The neuroscience of decision making is multidisciplinary, drawing from economics, psychology, and neuro-science (Gold and Shadlen, 2007; Lee, Seo, and Jung, 2012; Sanfey, Stallen, and Chang, 2014). For most psychologists, before making decisions people consider their internal state because decision making is a goal-directed behavior. Rangel, Camerer, and Montague (2008) developed a framework to highlight the processes that influence goal-directed choices. This framework is intended to build a biologically sound theory of how humans make decisions that can be applied in both the natural and the social sciences. Rangel and Hare (2010) explored how the brain processes computations of choices and valuations in decision making. Goal-directed or model-based decision approaches are used to pick the optimal action.

The following example illustrates a neural explanation of decision making as a goal-directed behavior. If an individual is looking for food, the internal state of hunger would have to be assessed first. After this assessment, the individual will engage in behavior oriented toward addressing the hunger. Very likely, getting food will allow this individual to solve this hunger issue. In this example, searching for food qualifies as a goal-directed behavior. Goal-directed behaviors tend to be influenced by the anterior cingulate cortex (ACC), the orbitofrontal cortex (OFC), and the ventromedial prefrontal cortex (vmPFC). The activation of these brain structures would motivate the person to look for food, which could be construed as a form of reward. "Undoubtedly, a large leap from precise neural activity to big decisions like planning for retirement or buying a car could resolve years or decades of debate that are difficult to resolve with other sorts of experiments" (Camerer and Loewenstein, 2004, p. 38). Bush and collaborators (2002) suggest that the "dorsal anterior cingulate cortex (dACC) may play a specific role in reward circuitry particularly in reward-based decision making, learning, and the performance of novel (non-automatic) tasks" (p. 527).

Decision making is supported by a distributed network of brain regions that includes the OFC, the ACC, the dorsolateral prefrontal cortex (dlPFC), the thalamus, the parietal cortices, and the caudate nucleus (Ernst and Paulus, 2005). The frontal lobes are involved in tasks from binary choices to making multi-attribute decisions that require explicit deliberation and integration of diverse sources of information (Krawczyk, 2002). Krawczyk (2002) divided the prefrontal cortex into three parts to account for its role in decision making: (1) the orbitofrontal and ventromedial areas are most relevant to deciding based on reward values and contribute to affective information regarding decision attributes and options; (2) the dorsolateral prefrontal cortex is critical in making decisions that call for the consideration of multiple sources of information; and (3) the anterior and ventral cingulate cortices appear especially relevant in sorting among conflicting options, as well as signaling outcome-relevant information.

Lesion studies have shown that damage to parts of the prefrontal cortex (PFC) impairs decision making. Kennerley and Walton (2011) found that patients with damage to the PFC, especially the ventral and medial parts, often show a marked inability to make choices that meet their needs and goals. Specifically, they found that patients with damage in these brain areas live "disorganized lives, tend to be impatient, vacillate when making decisions, often invest their money in risky ventures and exhibit socially inappropriate behavior" (Kennerley and Walton, 2011, p. 297). Three areas whose damage affects decisions are: the ACC, the OFC, and the ventromedial prefrontal cortex (vmPFC).

These impairments also affect the value computations that are necessary for optimal choice. These findings support previous studies by Damasio and collaborators who contend that the decision making process is influenced by marker signals that arise in bioregulatory processes, including those that express themselves in emotions and feelings (Bechara and Damasio, 2005). This influence can occur consciously or unconsciously. According to Bechara, Damasio, and Damasio (2000) and Bechara and Damasio (2005), somatic markers are automatic signals that indicate the positive and negative consequences of experienced stimuli and guide decision making. Decision making entails uncertainty and risk. In the following lines, I explore the neural foundations of these two concepts.

2 NEURAL BASIS OF UNCERTAINTY AND RISK

Decisions are often made in the presence of uncertainty about their outcomes. However, "people often prefer the known over the unknown, sometimes sacrificing potential rewards for the sake of surety" (Huettel et al., 2006, p. 765). Uncertainty refers to imperfect knowledge about how choices lead to outcomes (Platt and Huettel, 2008). For psychologists, "uncertainty refers to the psychological state in which a decision maker lacks knowledge about what outcome will follow from what choice" (Platt and Huettel, 2008, p. 398). Hence, uncertainty is inherent to decision making in organizations and in everyday life. Uncertainty can refer to risk, which is present when there are multiple possible outcomes that could occur with well-defined or estimable probabilities (Bernoulli [1738] 1954). However, a distinction between risk and uncertainty is often made in the economics literature.

Frank Knight (1921) was the first economist to differentiate uncertainty from risk. "Risk refers to situations where the decision maker knows with certainty the mathematical probabilities of possible outcomes of choice alternatives, whereas uncertainty refers to situations where the likelihood of different outcomes cannot be expressed with any mathematical precision" (Weber and Johnson, 2009, p. 130). In organizations, most decisions often have these two attributes. Managers and employees make decisions without clear knowledge of the potential outcomes. Under certain conditions, they may even not be aware of the potential implications of their decisions. Hence, one could argue that knowledge about the probability distribution of possible outcomes can range from complete ignorance to certainty (ibid.). Managers, for example, may completely be ignorant of the decision of a competitor to introduce a new product or lower prices but have clear knowledge of how many employees to lay off.

Take the example of hiring a new manager to head a department within an organization. The organization does not have a clear understanding of whether the new manager will succeed in the new position. Under these conditions, the organization relies on characteristics such as education, experience, reputation, and other attributes that could signal potential success (Spence, 1973). Once hired, the manager may or may not be successful. The new manager's success could be construed as a gain for the organization, whereas a failure could be construed as a loss. To reduce this uncertainty, the hiring process will use various means such as education, work experience, and background checks. Huettel et al. (2006) found that preferences for risk (uncertainty with known probabilities) and ambiguity (uncertainty with unknown probabilities) predict brain activation associated with decision making, and activation within the lateral prefrontal cortex was predicted by ambiguity preference, whereas activation in the posterior parietal cortex was predicted by risk preference.

Economists also make the distinction between uncertainty, risk, and ambiguity. In Knightian terms, uncertainty can also refer to ambiguity, in which there are multiple possible outcomes whose probabilities are not known or are not well defined. Ambiguity occurs when outcome probabilities are unknown. In general, people tend to be ambiguity averse (Camerer and Weber, 1992). A meta-analysis of decision making conducted by Krain et al. (2006) found that both risky and ambiguous situations lead to activation in the frontal and parietal regions, the thalamus, and the caudate nucleus. The authors also found that ambiguous decision making was associated with activity in the dorsolateral prefrontal cortex (dlPFC), regions of the dorsal and subcallosal anterior cingulate cortex (ACC), and parietal cortex, whereas risky decision making was associated with activity in orbitofrontal cortex (OFC), rostral portions of the ACC, and parietal cortex.

Risky decisions may involve uncertainty about possible outcomes (reward risks) or uncertainty about which action should be taken (behavioral risk). Both types of risk are activated by different regions of the brain. Huettel (2006) found that behavioral risk modulated activation in prefrontal, parietal, and insular regions. However, no effect of reward risk was observed in these brain regions. Reward risk reflects limited knowledge about which outcome will occur and represents a continuous quantity that is probabilistically expressed, whereas behavioral risk reflects limited knowledge about which potential action is the optimal one to choose and constitutes a categorical variable that is present or absent (ibid.). Scholars in behavioral finance, such as Knutson and Bossaerts (2007), invoke two opposing metrics: expected reward and risk when measuring the neural basis of financial decisions. Expected reward is the outcome that is anticipated and risk refers to the possibility of receiving (or not receiving) a positive outcome. They show that the ventral striatum plays a critical role in the representation of expected reward while the insula may play a more prominent role in the representation of expected risk.

Neuroscientists have suggested that organisms can make better decisions if they have at their disposal a representation of the uncertainty associated with task-relevant variables and the brain may use knowledge of uncertainty, confidence, and probability in making decisions (Ma and Jazayeri, 2014). Camerer et al. (2004) describe the amygdala as an internal "hypochondriac," which provides "quick and dirty" emotional signals in response to potential fears (p. 561). Several brain structures, such as the amygdala and the insula have been identified as playing a crucial role in uncertainty. These structures are also involved in emotional responses such as fear. The amygdala is a brain structure implicated in many kinds of phenomena such as attitudes, stereotyping, perception, and emotion (Ochsner and Lieberman, 2001). "The insula is a region that processes information from the nervous system about bodily states, such as physical pain, hunger, the pain of social exclusion, disgusting odors, and choking" (Camerer et al., 2004, p. 568). This suggests a neural basis for "fear of the unknown" that influences choices (Camerer et al., 2004).

Individual response to uncertainty may also depend on the type of uncertainty faced by the decision maker. To attest to this, Vartanian, Mandel, and Duncan (2011) studied the neural responses of 16 participants facing two types of uncertainty: uncertainty in the life and cash domains. Uncertainty in the life domain relates to decisions involving the prospect of saving lives (or not). Uncertainty in the cash domain relates to the prospect of earning cash (or not). Vartanian et al. (2011) observed that participants exhibited greater risk aversion, conflict, and sensitivity to negative feedback in the life domain, which they attribute to valuation of human lives. Supporting this assertion, choices to save lives activated the dorsal striatum, consistent with its role in context-sensitive reward processing. In contrast, choices to save cash activated the posterior insula, which they attributed to its role in probability signaling and risk prediction. These findings indicate that the valuation of human life could be related to moral considerations. In most cultures, people are socialized to value life more than material things. Such socialization could predispose people to value life rather than cash as observed in this study. Retting and Hastie (2003) also suggest that the prospect of saving lives may motivate decision makers more than the prospect of earning cash (Rettinger and Hastie, 2003).

A risky situation entails opportunities for gain or loss. Prospect theory shows that people are more sensitive to losses than gains (Kahneman and Tversky, 1979; Rick, 2011). Loss aversion refers to the tendency for losses to have greater hedonic impact than comparable gains (Rick, 2011). Loss aversion is so powerful that once people possess objects they are reluctant to get rid of them. According to Thaler (1980), people often demand much more to give up an object than they would be willing to pay to acquire it. He labelled this tendency the "endowment effect." This effect leads people to value an object more highly when they possess it than they would value the same object if they did not possess it (Thaler, 1980; Kahneman, Knetsch, and Thaler, 1990, 1991). Getting rid of an object that one owns creates a cognitive discomfort.

In exploring the neural basis of prospect theory (Kahneman and Tversky, 1979), one might contemplate the following questions. Are losses and gains asymmetrically treated in the brain? If so what are the brain regions that process information related to losses and gains? Evidence in neuroeconomics indicates that people develop a propensity to experience direct pain when they spend money (Camerer et al., 2004; McClure, York, and Montague, 2004). Yang et al. (2007) focused on high-conflict conditions (probability of losing about 50 percent) and low-conflict conditions (probability of losing about 20 percent) and found that high-conflict conditions elicited more negative event-related potential deflections than low-conflict conditions. Hence, the prospect of losing a large amount of money is more painful in the brain than the prospect of losing a lesser amount.

Compared to gain, loss is considered painful. Consequently, enhanced sensitivity to losses is driven by emotions such as fear and anxiety. The neural substrates of such emotions include the amygdala and the insula (Tom et al., 2007). In organizations, most decisions involve the possibility of gaining or losing something. For example, a merger or acquisition decision involves the possibility for the focal firm to gain or lose. The same is true whether an employee accepts a promotion or not. Loss aversion could also reflect an asymmetric response to losses versus gains within a single system that codes for the subjective value of the potential gamble, such as the ventromedial prefrontal cortex, the orbitofrontal cortex, and the ventral striatum. As described above, this was evidenced by Huettel et al. (2006) who found that preferences for risk (uncertainty with known probabilities) and ambiguity (uncertainty with unknown probabilities) predict brain activation associated with decision making. Activation within the lateral prefrontal cortex was predicted by ambiguity preference, whereas activation in the posterior parietal cortex was predicted by risk preference.

Research also indicates that earned money is more rewarding in the brain than unearned money (Camerer et al., 2004). These findings may have direct implications for managers. In organizations, tying compensation to performance may be more rewarding for employees than compensation that is not performance based. Dickaut et al. (2003) found more activity in the orbitofrontal cortex when thinking about gains compared to losses, and more activity in inferior parietal and cerebellar areas when thinking about losses. They also found that when people experience pleasure or anticipate pleasure, the nucleus accumbens is activated and the insula is activated when people experience or anticipate pain, taste something bad, or see a disgusting picture. Several authors have observed that people are more averse to ambiguity than to risk alone (Platt and Huettel, 2008, p. 398). Using functional magnetic resonance imaging (fMRI), Hsu et al. (2005) found that the level of ambiguity in choices correlated positively with activation in the amygdala and the orbitofrontal cortex, and negatively with the striatum system. They also found that striatal activity correlated positively with expected reward.

3 NEURAL FOUNDATIONS OF VALUE-BASED DECISIONS

Making choices implies that one assigns values to particular outcomes. According to Bernoulli ([1738] 1954), the determination of the value of an item is not based on its price but on the utility it yields. For example, a sum of $1000 may not have the same value for a rich person as for a poor person. A poor person may consider a reward of such amount as an extremely valuable windfall, whereas a rich person may perceive it as a token. When faced with choices, people tend to develop a cognitive map of the options available and the valuation of these options (O'Doherty, Cockburn, and Pauli, 2016). The valuation process has neural underpinnings. Kennerley and Walton (2011) note that "many neurons in the brain are modulated by the value of an outcome" (p. 298). What then are the brain structures that assign value to potential outcomes? O'Doherty et al. (2016, p. 83) note that the "regions of the frontal and parietal cortices play a fundamental role in the computation of model-based action values."

Humans tend to view choice options not in absolute terms but rather as relative to salient reference points (Kahneman, 2011). The decision maker assigns values to different alternatives. In terms of understanding the role of the brain in value-based decisions, the question becomes: how does the brain decide to attribute value to some options and ignore others

(Rangel et al., 2008)? For example, in organizations, employees assign positive values to outcomes such as promotion, bonuses, raises, and good performance evaluations, and negative values to outcomes such as reprimands, demotion, firing, and poor performance evaluations. Although management research indicates that employees strive to receive positive outcomes and avoid negative ones, it is not clear how the brain processes such decisions. Because the human brain works by association, people learn to associate positive values with some outcomes and negative values with others.

Behaviors in organizations are forms of social behavior and therefore guided by social expectations. For example, employees are expected to behave in certain ways within organizations; failure to do so would amount to violations of organizational norms, which are examples of social norms. How then do managers and other employees react to violations of such organizational norms? What can we learn from neuroscience when exploring reactions to such violations? Neuroscientific research found that the anterior cingulate cortex processes the tracking of social expectation violations, hence extending previous computational conceptualizations of this region to the social domain (Chang and Sanfey, 2013).

According to Rangel and Hare (2010), models of decision making in economics and psychology contend that the brain computes stimulus values that measure the value of the outcomes generated by each action. It then computes action costs that measure the costs associated with each course of action and integrates them into action values given – hence the following formula: Action Value = Stimulus Value – Action Cost. Finally, the action values are compared in order to make a choice. The brain encodes stimulus values through activation of the medial orbitofrontal cortex (mOFC).

Camerer and Mobbs (2017) compared brain activity during hypothetical and real choices and found that brain activity was more intense and widespread during real choices than hypothetical ones. The authors found that in many cases hypothetical choice tasks give an incomplete picture of the brain circuitry that is active during real choice. There are typically differences in the intensity of neural activation when subjects make real versus hypothetical choices. In addition, there are often distinct patterns of neural activation when subjects make real choices, presumably reflecting distinct neural mechanisms that are only engaged by real choices. The implication is that studies based on hypothetical choice can give an incomplete picture of brain activity during real choice. For this reason, it would be valuable to have more studies comparing hypothetical and real choice within a common paradigm: "Then proper meta-analysis can be done, and provide guidance about when hypothetical choice gives the most incomplete picture" (Camerer and Mobbs, 2017, p. 54). One of the limitations of this study is that experimental subjects often make hypothetical choices with no direct consequences to them. Hence, it could be difficult to generalize such findings to real-life situations because when faced with real stimuli, people may display behaviors that could be more intense than those they displayed during the hypothetical scenarios. Seeing the picture of a frightening animal may not lead to the same intensity of emotion as actually facing the same animal in a real-life situation.

Tobler et al. (2007) used fMRI to identify brain activations coding the key decision parameters of expected value (magnitude and probability) separately from uncertainty (statistical variance) of monetary rewards. Participants discriminated behaviorally between stimuli associated with different expected values and uncertainty. Stimuli associated with higher expected values elicited monotonically increasing activations in distinct regions of the striatum, irrespective of different combinations of magnitude and probability. Stimuli associated with higher uncertainty (variance) activated the lateral orbitofrontal cortex. Uncertainty-related activations covaried with individual risk aversion in lateral orbitofrontal regions and risk seeking in more medial areas.

Furthermore, activations in expected value-coding regions in prefrontal cortex covaried differentially with uncertainty depending on risk attitudes of individual participants, suggesting that separate prefrontal regions are involved in risk aversion and seeking. Using a Bayesian reinforcement-learning model and fMRI, Boorman et al. (2009) show that the frontopolar cortex (FPC) tracks the relative advantage in favor of switching to a foregone alternative when choices are made voluntarily. Changes in FPC functional connectivity occur when subjects finally decide to switch to the alternative behavior. Moreover, interindividual variation in the FPC signal predicts interindividual differences in effectively adapting behavior. By contrast, the ventromedial prefrontal cortex (vmPFC) encodes the relative value of the current decision.

The tendency to devalue rewards that are offered in the future is known as time discounting. Using event-related fMRI on a sample of 41 healthy 18–24-year-old males, Waegeman et al. (2014) found that choosing the delayed option activated the inferior frontal gyrus, lateral and ventrolateral prefrontal cortices, and the lateral orbitofrontal cortex. Decisions of individuals who delay more often during the task are associated with more activity in the dorsolateral prefrontal cortex. They furthermore show correlated activity between the inferior frontal gyrus, dorsolateral prefrontal cortex and medial prefrontal regions during decision making compared with individuals who behave more impulsively. The authors also found that choosing the earlier reward was not associated with any increases in brain activation compared to choosing the delayed reward. However, individuals who behave more impulsively show more activation in the medial prefrontal cortex (anterior cingulate cortex, medial frontal gyrus), and no correlated activity with the inferior frontal gyrus. Waegeman et al. (2014) conclude that individual differences in self-control during time discounting may partly result from differential activation of the dorsolateral prefrontal cortex.

Even decision making aspects, such as the framing effect, have neural foundations. How people frame problems affects the way they approach and solve them (Tversky and Kahneman, 1981, 1986; Kahneman and Tversky, 1984). De Martino et al. (2006) found that the framing effect was associated with activity in the amygdala, which indicates a key role of the emotional system. However, orbital and medial prefrontal cortex activity predicted a reduced susceptibility to the framing effect.

Human decisions are also guided by outcomes that are associated with decisions made in the past. Behrens et al. (2007) found that participants assess volatility in an optimal manner and adjust decision making accordingly. This optimal estimate of volatility is reflected in the fMRI signal in the anterior cingulate cortex when each trial outcome is observed. When a new piece of information is witnessed, activity levels reflect its salience for predicting future outcomes (ibid.). Bayesian analysis suggests that optimal learning for decision making should reflect the salience of each new piece of information for predicting future outcomes.

The neural basis of other decision making models, such as the diffusion decision model (DDM), has been explored. The DDM helps to assess the speed and accuracy of decisions. Sequential sampling models assume that people make speedy decisions by gradually accumulating noisy information until a threshold of evidence is reached. Most real-life decisions are composed of two separate decisions: (1) the decision to stop deliberating; and (2) the decision or act itself (Forstmann, Ratcliff, and Wagenmakers, 2016, p. 642). As Forstmann et al. (2016, p. 644) note, "the sequential nature of decision making is a fundamental property of the human nervous system, reflecting its inability to process information instantaneously." Most decisions are made under time pressure. From routine to deliberative decisions, people cannot spend time pondering over what choice to make. At some point, they have to make a choice. "People often need to make decisions based on information that unfolds over time" (Forstmann et al., 2016, p. 643). Making decisions in this manner often occurs in organizations. Managers and employees often process information sequentially.

We must acknowledge that organizations are complex systems that are themselves embedded in other complex systems, such as communities, nations, and the entire global societal system. Hence, the decisions humans make are often made in the context of social interactions. As a result, it could be difficult to link the choices we make to the activity of a single neuron. Santos and Rosati (2015) reviewed the evolutionary basis of human decision making and concluded that non-human species displayed the same types of decision biases and heuristics that humans show. Merely recognizing a bias does not make this bias go away.

4 ORGANIZATIONAL IMPLICATIONS

The context in which decisions are made is important in analyzing and understanding the decision making process. This has led several authors to suggest that the field of judgment and decision making has reached a stage in which context dependence must be seen as central to theory, as something that cannot be ignored without incurring a severe loss of explanatory completeness (Goldstein and Weber, 1997). In describing the job of a manager, Mintzberg (1979) identified the manager's decision roles. Managers make decisions covering aspects such as hiring, firing, promoting, which strategy to pursue, which market to target, and whether to enter into agreements with other firms. Although the rational choice model may contend that such decisions are rational, understanding of psychology, neuroscience, and neuroeconomics indicates that such decisions are far from rational. In fact, such decisions may have a neural basis. Moreover, using neuroscientific tools could help better explain decisions made in organizations.

Although decision making is a cognitive process, in studies of patients with brain lesions (Damasio, 1994; Bechara et al., 2000; Bechara and Damasio, 2005) it is shown that emotions affect and are an integral part of the decision making process. Patients who do not experience emotions are impaired in their ability to make decisions.

The assumption that "the goals of a decision maker are to achieve desired outcomes and avoid undesired ones" (Gold and Shadlen, 2007, p. 538) is consistent with the avoidance response in psychology. In the workplace, this could translate into positive experiences such as social and monetary rewards. Employees seek such rewards, whereas they avoid unpleasurable experiences such as punishment, abusive supervision and undeserved criticism. "To select appropriate behaviors leading to rewards, the brain needs to learn associations among sensory stimuli, selected behaviors, and rewards" (Haruno and Kawato, 2006, p. 948). Haruno and

Kawato (2006) conducted an event-related fMRI study and found that neural correlates of the stimulus–action–reward association reside in the putamen, whereas a correlation with reward-prediction error was found largely in the caudate nucleus and the ventral striatum. In organizations, employees make decisions to seek preferred outcomes and avoid unpleasant ones. For example, an employee who desires to earn a promotion would exercise the necessary effort to accomplish his or her goal. Likewise, an employee may arrive on time or work hard to avoid dismissal.

Laureiro-Martinez et al. (2014) compared entrepreneurs and managers on the exploration–exploitation dichotomy and found that compared with managers, entrepreneurs showed higher decision making efficiency and a stronger activation in regions of the frontopolar cortex (FPC) previously associated with explorative choice. Decision efficiency was measured as performance by response time. An optimal balance between efficient exploitation of available resources and creative exploration of alternatives is critical for adaptation and survival. Adaptive behavior in an uncertain world requires managing the trade-off between exploiting known sources of reward and exploring the environment to gather information about different, potentially more valuable options. Daw et al. (2006) found that explorative choices engage the intraparietal sulcus and lateral prefrontal regions, particularly the FPC. Exploitative choices also engage the dopaminergic frontolimbic striatal system projecting to the ventromedial prefrontal cortex (Beeler et al., 2010).

4.1 Automatic versus Deliberate Decision Making

Neurocognitive psychologists (Satpute and Lieberman, 2006; Lieberman, 2007a, 2007b) and behavioral economists (Kahneman, 2003, 2011) have identified two cognitive systems in decision making: System 1 and System 2.

System 1 refers to automatic processing of information often based on intuition and experience. System 1 corresponds to the X-system. The X-system is construed as "a pattern matching system whose connectivity weights are determined by experience and whose activation levels are determined by current goals and features of the stimulus input" (Lieberman et al., 2002, p. 211). System 2 is more elaborate and effortful; it requires the use of deliberate reasoning, and corresponds to the C-system described by Lieberman and colleagues (2002). In organizations, System 1 decisions relate to those made based on experience and existing routines. For example, decisions related to how employees handle their daily job routines, such as inspecting equipment, scheduling tasks, or attending meetings, follow standardized processes and would be considered as automatic and require little cognitive effort. However, decisions related to planning, executing, developing new products, and targeting new markets would require more analytical effort. Such decisions would call on System 2.

Research has shown that the two systems activate different regions of the brain. System 1 activates the limbic system, whereas System 2 decisions activate the neocortex. The existence of these two systems indicates that the choices organizational members make are not always rational. Using these two systems could shed light on understanding certain decisions made in organizations. Simon (1965) made the distinction between programmed and non-programmed decisions. The former are routine, repetitive, and from the perspective of cognitive neuroscience, could recruit System 1. Non-programmed decisions, however, are often novel and unfamiliar. They require more cognitive effort than programmed decisions and are part of System 2. Administrative decisions are well programmed, whereas strategic decisions are non-programmed decisions.

To some extent, Simon's classification is closer to that of Ansoff (1965), who identified three types of managerial decisions: (1) administrative decisions; (2) operating decisions; and (3) strategic decisions. Administrative decisions are often routine and repetitive. Strategic decisions tend to be novel. Operating decisions often show overlap between elements of administrative and strategic decisions. Administrative decisions are made by lower-level managers, whereas operating decisions are made by middle-level managers, and strategic decisions by top-level managers.

Novel stimuli require more cognitive effort, whereas familiar situations require less. We know that experience and practice may lead a rather complex behavior to become automatic. As Camerer et al. (2004) note: "When good performance becomes automatic (in the form of procedural knowledge), it is typically hard to articulate, which means human capital of this sort is difficult to reproduce by teaching others" (p. 560). It also becomes part of what is known as tacit or implicit knowledge. Implicit knowledge is not codified and can only be transferred through interactions.

Automaticity may also help to explain the extent to which it is difficult to share tacit knowledge in organizations. It may also account for the role of experience in learning and performance because the existence of prototypes may play an important role in this process. People compare new events to prototypes already stored in memory. In explaining how entrepreneurs recognize opportunities for new ventures, Baron (2006) notes that experienced, serial entrepreneurs generally search for opportunities in areas or industries where they are already knowledgeable. Such a comparison helps individuals to make decisions fast and effortlessly in familiar situations, whereas novel ones require more cognitive effort.

4.2 Exploitation–Exploration Principles in Decision Making

A Nightingale, sitting aloft upon an oak, was seen by a Hawk, who made a swoop down, and seized him. The Nightingale earnestly besought the Hawk to let him go, saying that he was not big enough to satisfy the hunger of a Hawk, who ought to pursue the larger birds. The Hawk said: "I should indeed have lost my senses if I should let go food ready to my hand, for the sake of pursuing birds that are not yet even within my sight." (*Aesop's Fables. A New Revised Version From Original Sources*, 1884, cited in Bunge and Wendelken, p. 609)

What then would lead an individual to switch from exploiting a current opportunity at hand to look for another one? What brain mechanism can lead to the switch of behavior? As the fable indicates, a bird in the hand is worth two in the bush unless the probability of catching the birds in the bush is very high. "The evidence favoring an alternative choice is tracked by the lateral frontopolar cortex and this information appears to be transmitted to the inferior parietal sulcus area and ventral premotor cortex in advance of a switch behavior" (Bunge and Wendelken, 2009, p. 609). The ventromedial prefrontal cortex encodes the immediate relative value of the current choice. Using fMRI, Daw et al. (2006) showed that the frontopolar cortex and intraparietal sulcus are preferentially active during exploratory decisions. In contrast, regions of striatum and ventromedial prefrontal cortex exhibit activity characteristic of an involvement in value-based exploitative decision making.

Organizational scholars refer to the constructs of exploitation and exploration to explain the dilemma described in the fable (March, 1991). Exploration entails disengaging from the current task to enable experimentation, flexibility, discovery, and innovation. Exploitation aims at optimizing the performance of a certain task and is associated with high-level engagement, selection, refinement, choice, production, and efficiency. Exploitation is defined as behavior that optimizes performance in the current task, whereas exploration is the "disengagement from the current task and the search for alternative behaviors" (Aston-Jones and Cohen, 2005, p. 403). "The essence of exploitation is the refinement and extension of existing competences, technologies, and paradigms. Its returns are positive, proximate, and predictable," whereas "the essence of exploration is experimentation with new alternatives. Its returns are uncertain, distant, and often negative" (March, 1991, p. 85). People switch from exploitation to exploration and vice versa. People tend to shift from the current task when the utility of performing it falls under a certain threshold. Maintaining an appropriate balance between exploration and exploitation is a primary factor in system survival and prosperity (March, 1991). The trade-off between exploitation and exploration represents a challenge to behavior at all levels and over multiple time scales (Cohen, McClure, and Yu, 2007).

Laureiro-Martinez et al. (2015) use fMRI in a sample of expert decision makers and found that exploitation activates regions associated with reward seeking that track and evaluate the value of current choices, while exploration relies on regions associated with attentional control that track the value of alternative choices. Exploitation involves bottom-up learning. The dorsolateral prefrontal cortex and the frontopolar cortex are responsible for top-down control over attention. Laureiro-Martinez et al. (2015) found that exploitation relies on brain regions mainly associated with anticipation of rewards, whereas exploration depends on regions mainly associated with attentional control: "Sustained high performance depends, not on individual specialization, but on the ability to shift between exploitation and exploration, which in turn depends on stronger activation of the brain regions responsible for attentional and cognitive control" (p. 333).

Laureiro-Martinez et al. (2015) also note that exploration has an emotional cost because it involves abandoning less uncertain gains for more uncertain but potentially larger rewards. According to Laureiro-Martinez, Brusoni, and Zollo (2010), adaptive firm behavior in a diverse and rapidly changing environment requires a trade-off between exploiting known sources of reward and exploring the environment for more valuable or stable opportunities (p. 95). At the individual level of analysis, exploration and exploitation are orthogonal, while at the group level some members can pursue exploitation and others can focus on exploration. Hence, exploration and exploitation can be on a continuum at the group level of analysis.

4. The neural basis of creativity and innovation

Creativity is essential to the development of human civilization and plays a crucial role in cultural life (Takeuchi et al., 2010). As Dietrich and Kanso (2010) put it, "creativity is the fountainhead of human civilizations. All progress and innovation depend on our ability to change existing thinking patterns, break with the present, and build something new" (p. 822). Hence, its study must be seen as a basic necessity (Hennessey and Amabile, 2010). Creativity helps people to adapt to changing circumstances in their environment. This is particularly true for organizations that not only need to adapt to constant changes in their external environment but also must anticipate the potential impact of those changes. Hence, creativity is more important now than ever before and can be viewed as "a useful and effective response to evolutionary changes" (Runco, 2004, p. 658).

Psychologists and neuroscientists have long explored the sources of creativity and genius: Abraham et al. (2012, p. 1906) contend that, "our fundamental capacity to be creative is a subject of much fascination to scientists and lay people alike." What led people such as Pablo Picasso, Shakespeare, Mozart, Einstein and many others who have changed the course of human history to make their remarkable discoveries? Where did their ideas come from? Could neuroscience help us address some of the sources of human creativity? And can such understanding help organizations create environments that are conducive to creativity and innovation? Answers to these questions have garnered the interest of psychologists, organizational scholars, and neuroscientists. The aim of a neurocognitive perspective of creativity is to uncover the neural substrates of the cognitive processes that substantiate creative thinking.

1 UNDERSTANDING THE CREATIVE PROCESS

1.1 Defining Creativity

Although psychologists have studied creativity for more than 60 years, its definition still remains unsettled. However, most researchers agree that creativity is the ability to produce work that is novel, useful, and generative (Stein, 1953; Mednick, 1962; Amabile, 1996; Sternberg and Lubart, 1996; Runco and Jaeger, 2012). The two important aspects of creativity are novelty and usefulness. Indeed, "a creative idea is one that is novel, surprising and valuable" (Boden, 1998, p. 347). According to Runco and Jaeger (2012), "if something is not unusual, novel, or unique, it is common place, mundane, or conventional. It is not original, therefore, not creative" (p. 92). However, the authors acknowledge that originality is necessary but not sufficient for an idea to be considered creative; something creative must also be effective (or useful). "Edison's invention of the electric light may not be an example of extraordinary complex creation, but it is valued and remembered by its usefulness" (Welling, 2007, p. 164).

Creativity is a multifaceted ability (Ward, 2007; Pieritz et al., 2012) and involves the ability to understand and express novel, orderly relationships (Caselli, 2002; Chakravarty, 2010). It is based on the processing of remote or loose associations between ideas (Mednick, 1962). Defined as such, creativity is neither a special "faculty" nor a psychological property confined to a tiny elite. Rather, it is a feature of human intelligence in general and is "grounded in everyday capacities such as the association of ideas, reminding, perception, analogical thinking, searching a structured problem-space, and reflective self-criticism" (Boden, 1998, p. 347). "Creativity involves not only a cognitive dimension (the generation of new ideas) but also motivation and emotion, and is closely linked to cultural context and personality factors" (ibid.). Takeuchi et al. (2010, p. 14) consider creativity as a complex cognitive function that requires diverse cognitive abilities, such as working memory, sustained attention, cognitive flexibility, and fluency in the generation of ideas and in the judgment of propriety, which are typically ascribed to the prefrontal cortex.

1.2 The Creative Thinking Process

The creative process has been described as a series of cognitive stages that culminate in the generation of a novel and useful idea. The phases of the process include: (1) preparation; (2) incubation; (3) illumination; and

(4) verification (Wallas, 1926). In the first phase, preparation, the individual is obsessed with the idea and collects information about it. In the second stage, the individual does not actively attempt to solve the problem; however, he or she unconsciously continues to work on it. In the third stage, a possible solution surfaces to consciousness. The solution may not be well defined at this stage. Finally, in the fourth stage, the idea is worked into a form that can be proven and communicated to others. In this process, conceptual expansion, that is, the constraining influence of examples, creative imagery and insight are important components (Abraham and Windmann, 2007).

Mednik (1962) defined the creative thinking process as the forming of associative elements into new combinations that either meet specified requirements or are in some way useful (p. 221). Mednik also identified three ways to bring about a creative solution: (1) serendipity; (2) similarity; or (3) mediation. Serendipity refers to the discovery of a novel idea by accident. In science, the arts, and other disciplines, there are multiple examples of serendipity. Mednick (1962) cites the invention of the X-ray and the discovery of penicillin as examples of serendipity. Similarity involves association between elements that may be apparently dissimilar. "This mode of creative solution may be encountered in creative writing which exploits homonymity, rhyme, and similarities in the structure and rhythm of words or similarities in the objects which they designate" (p. 222). In mediation, the "requisite associative elements may be evoked in contiguity through the mediation of common elements" (ibid.). Construing creativity as a process entails considering it as a sequence of cognitive operations that "creative" individuals perform and that lead to the creation of novel and useful ideas.

Creativity is considered as both a process and an output. As a process, creativity includes a series of cognitive steps that lead to the generation of novel and useful ideas. As an outcome, it focuses on the quality of the idea generated – whether it is novel and useful.

1.3   Creative Cognition

The creative cognition approach is rooted in cognitive psychology and cognitive science and "assumes that the cognitive capacity to behave creatively is a normative characteristic of humans, and it seeks to advance our understanding of creativity through precise characterization and rigorous scientific study of the cognitive processes that lead to creative and noncreative outcomes" (Ward, 2007, pp. 28–29). "Creative cognition can be understood as a set of cognitive processes that support the generation of novel and useful ideas" (Beaty et al., 2016, p. 87). It is the set of cognitive processes and metacognitive strategies used during creative production (Li et al., 2015). Several scholars have discussed the role of creative cognition concepts, such as divergent thinking, working memory, and cognitive load in the generation of novel ideas. One creative cognition that has been studied and often equated to creativity is divergent thinking. Divergent thinking is the ability to generate multiple solutions to an open-ended problem (Guilford, 1967). It "involves the ability to consciously generate new ideas that branch out and allow for many possible solutions to a given problem" (Shamay-Tsoory et al., 2011, p. 178).

"Creative problem solving as opposed to analytical problem solving, does not involve computational algorithms or incremental analytical procedures. Instead, creative problem solving tends to be characterized by more divergent, associational or discontinuous solution processes" (Jarosz, Colflesh, and Wiley, 2012, p. 487). In the study of creativity and problem solving, two general strategies have been considered: a methodical, conscious search of problem-state transformations, and a sudden insight with abrupt emergence of the solution into consciousness (Kounios et al., 2008). Insight occurs when problem solutions arise suddenly and seem obviously correct, and is associated with an "Aha!" experience (Kounios et al., 2006).

Because the brain is involved in most behaviors displayed by humans, "any theory of creativity must be consistent and integrated with contemporary understanding of brain function" (Pfenninger and Shubik, 2001, p. 217). This is particularly important because little is known about the brain mechanisms that underlie creative thinking (Dietrich, 2004a). However, recent research has used neuroscientific methods to study originality, novelty, insight, divergent thinking, and other processes related to creative mental activity (Yoruk and Runco, 2014). These findings indicate that both hemispheres are involved in divergent thinking, which is accompanied by both event-related increases and decreases in neural activation.

2 NEURAL BASIS OF CREATIVITY

2.1 Brain Structures Involved in Creativity

Early studies on the neural basis of creativity led to the emergence of two distinct views on the neural basis of creative thinking: "1) the faculties for creative cognition reside predominately in the right hemisphere and 2) creative cognition is derived from reduction in network inhibition, which originates in the frontal lobe" (Kaufman et al., 2010, p. 217). Abraham et al. (2012) and Pieritz et al. (2012) found that brain regions involved in the retention, retrieval, and integration of conceptual knowledge such as the anterior inferior frontal gyrus, the temporal poles, and the lateral frontopolar cortex were selectively involved during conceptual expansion. Likewise, Takeuchi et al. (2010) found that the frontal lobe and the striatum are associated with cognitive flexibility and with creativity. These findings go against generic ideas that argue for the dominance of the right hemisphere during creative thinking and indicate the need to reconsider the functions of regions such as the anterior cingulate cortex to include more abstract facets of cognitive control.

Further studies by Kounios et al. (2006) indicate that neural activity during a preparatory interval before participants saw verbal problems predicted which problems they would subsequently solve with, versus without, self-reported insight. Specifically, electroencephalographic (EEG) topography and frequency (Experiment 1) and functional magnetic resonance imaging (fMRI) signals (Experiment 2) suggest that mental preparation leading to insight involves heightened activity in medial frontal areas associated with cognitive control and in temporal areas associated with semantic processing. The results for EEG topography suggest that non-insight preparation, in contrast, involves increased occipital activity consistent with an increase in externally directed visual attention.

Chermahini and Hommel (2010) studied whether individual performance in divergent thinking (alternative uses task) and convergent thinking (Remote Association Test, RAT) can be predicted by an individual's spontaneous eye blink rate (EBR), a clinical marker of dopaminergic functioning. The authors found that EBR predicted flexibility in divergent thinking and convergent thinking, but in different ways. The relationship with flexibility was independent of intelligence and followed an inverted U-shaped function, with medium EBR being associated with greater flexibility. Convergent thinking was positively correlated with intelligence but negatively correlated with EBR, suggesting that higher dopamine levels impair convergent thinking. Divergent thinking benefitted most from medium EBRs, while convergent thinking was best with low EBRs.

Behavioral research indicates four salient features of insightful problem solving: (1) mental impasse; (2) restructuring of the problem representation; (3) a deeper understanding of the problem; and (4) an "Aha!" feeling of suddenness and obviousness of the solution (Sandkühler and Bhattacharya, 2008). Sandkühler and Bhattacharya (2008) also observed an increased upper alpha band response in right temporal regions (suggesting active suppression of weakly activated solution-relevant information) for initially unsuccessful trials that after hints led to a correct solution.

Kaplan and Simon (1990) found that noticing properties of the situation that remained invariant during solution attempts (the notice invariants heuristic) proved to be a particularly powerful means for focusing search. In conjunction with hints and independently, the notice invariants heuristic played a major part in producing the insight that yielded an effective problem representation and solution. An insight is generally defined as a sudden comprehension that solves a problem, reinterprets a situation, explains a joke, or resolves an ambiguous percept (Sternberg and Davidson, 1995). Divergent thinking seems to be associated with high neural activation in the central, temporal, and parietal regions, indications of semantic processing and recombination of semantically related information (Yoruk and Runco, 2014).

Wu et al. (2015) conducted a meta-analysis of fMRI studies of divergent thinking and found that distributed brain regions were more active under divergent thinking tasks than those under control tasks, but a large portion of the brain regions were deactivated. The brain networks of the creative idea generation in divergent thinking tasks (DTTs) may be composed of the lateral prefrontal cortex, posterior parietal cortex, anterior cingulate cortex (ACC), and several regions in the temporal cortex, and the left fusiform gyrus related to selecting the loosely and remotely associated concepts and organizing them into creative ideas, whereas the ACC was related to observing and forming distant semantic associations in performing DTTs.

Aziz-Zadeh, Liew, and Dandekar (2013) used fMRI to measure neural activity in participants solving a visuospatial creativity problem that involved divergent thinking and was considered a canonical right hemisphere task. They found that both the visual creativity task and the control task activated the posterior parietal cortex and motor regions, which are known to be involved in visuospatial rotation of objects. Directly comparing the two tasks indicated that the creative task more strongly activated left hemisphere regions including the posterior parietal cortex, the premotor cortex, the dorsolateral prefrontal cortex (dlPFC) and the medial prefrontal cortex (mPFC). These results indicate that even in a task that is specialized to the right hemisphere, robust parallel activity in the left hemisphere supports creative processing. Furthermore, the results support the notion that higher motor planning may be a general component of creative improvisation and that such goal-directed planning of novel solutions may be organized top-down by the left dlPFC and by working memory processing in the medial prefrontal cortex.

Lesion studies have also been conducted to explore the neural basis of creativity. For example, Shamay-Tsoory et al. (2011) found that lesions in the mPFC involved the most profound impairment in originality. Furthermore, precise anatomical mapping of lesions indicated that while the extent of lesion in the right mPFC was associated with impaired originality, lesions in the left posterior parietal cortex (PC) were associated with somewhat elevated levels of originality. The authors note that a positive correlation between creativity scores and left PC lesions indicated that the larger the lesion is in this area, the greater the originality. On the other hand, a negative correlation was observed between originality scores and lesions in the right mPFC.

2.2 Creativity and the Dual System

The literature in social cognitive neuroscience identifies two types of systems that govern human cognition. These systems have been labeled as either the X-system and C-system (Lieberman et al., 2002; Lieberman, 2007a, 2007b), System 1 and System 2 (Kahneman, 2003, 2011; Camerer, Loewenstein, and Prelec, 2005) or implicit and explicit processes (Moors and De Houwer, 2010). According to Lieberman et al. (2002, p. 211), the X-system is a "pattern-matching system whose connectivity weights are determined by experience and whose activation levels are determined by current goals and features of the stimulus input." As such, it depends on the associative links formed through extensive learning histories, thereby leading to the automaticity of cognitive processes (Lieberman et al., 2002). One of the main characteristics of this system is effortless information processing (Dietrich, 2004b). Automatic processes recruit brain structures, such as the amygdala, the basal ganglia, the dorsal anterior cingulate cortex, the lateral temporal cortex, and the ventromedial prefrontal cortex (Lieberman et al., 2002).

Lieberman et al. (2002, p. 204) describe the C-system as "a serial system that uses symbolic logic to produce the conscious thoughts that we experience as reflections on the stream of consciousness." It involves systematic and deliberate action and requires cognitive effort. Its neuroanatomy includes the lateral prefrontal cortex, the posterior parietal cortex, the medial prefrontal cortex, the rostral anterior cingulate cortex, and the hippocampus, and surrounding medial temporal lobe region (Lieberman et al., 2002). The explicit system is associated with the higher cognitive functions of the frontal lobe and medial temporal lobe structures and has evolved to increase cognitive flexibility. In contrast, the implicit system is associated with the skill-based knowledge supported primarily by the basal ganglia and has the advantage of being more efficient (Dietrich, 2004a, 2004b).

Researchers studying creative cognitions (Evans, 2008, 2009; Allen and Thomas, 2011; Sowden, Pringle, and Gabaro, 2015) have also adopted this dual approach, which describes the creative process as involving two cognitive processes: Type 1 and Type 2. Type 1 processes are rapid, unconscious, automatic and intuitive. Type 2 processes, however, are normative and rational. This conceptualization begs the question of whether the dual system of thinking could improve our understanding of the neural basis of creativity. For example, Beaty et al. (2014) note that creative thought is strategic and controlled. This view is consistent with the controlled-attention theory of creativity, which contends that creative thought is a top-down process that taps into individual differences in the ability to control attention and cognition (ibid.). However, the associative theory of creativity contends that creative thought is unconscious and associative, which is consistent with the X-system or System 1.

The neuroscientific study of creativity shows that insight is the culmination of a series of brain states and processes operating at different time scales. Elucidation of these precursors suggests interventional opportunities for the facilitation of insight. According to Kounios and Beeman (2009), the brain response associated with the "Aha" moment was the culmination of a series of neural events, such as the alpha "brain blink" (p. 212). The brain regions often affected by insight include the right anterior temporal lobe, the right anterior superior temporal gyrus, and the right occipital cortex.

"When a weakly activated problem solution is present in the right temporal lobe, a temporary reduction in interfering visual inputs facilitates the retrieval of this solution, allowing the solution to pop into awareness" (Kounios and Beeman, 2009, p. 212). Jausovec (2000) found that creative people have more inter- and intrahemispheric EEG coherence than those who are less creative during an open-ended task (essay writing). Creative inspiration occurs in a mental state where attention is defocused, when thought is associative, and when a large number of mental representations are simultaneously activated (Martindale, 1999). This is characteristic of the default mode network, which is involved in complex, evaluative, and unconscious forms of information processing and contrasts with the cognitive control network, which tends to be more systematic, controlled, and strategic (Cole and Schneider, 2007). The following section discusses the role of the default mode and executive control networks in creativity.

2.3   Default Mode and Executive Control Networks and Creativity

Two neural networks, the default mode network (DMN) and the executive control network (ECN) seem to play a critical role in creativity. The default network represents a group of neurons that are active when the brain is at rest. These neurons imply that the brain never rests. The default mode network includes the medial prefrontal cortex (mPFC), the posterior cingulate cortex (PCC)/precuneus and the temporoparietal junction (TPJ) (Ellamil et al., 2012). It also includes midline and posterior inferior parietal regions that show increased metabolic activity in the absence of most externally presented cognitive tasks (Raichle et al., 2001).

Jung et al. (2013) note that the default mode network may provide a first approximation regarding how creative cognition might map onto the human brain. Focus on cortical hubs within the DMN represents a research opportunity to further refine the manifestation of creative cognition in the brain. Several authors (Raichle et al., 2001; Buckner, Andrews-Hanna, and Schacter, 2008; Andrews-Hanna, Smallwood, and Spreng, 2014) suggest that the default mode network activity is associated with spontaneous and self-generated thought, such as autobiographical retrieval, episodic future thinking, mental simulation, mental wandering, and social cognition. On the other hand, scholars have indicated the role of the executive control network in creative thinking.

The control network comprises lateral prefrontal and anterior inferior parietal regions and its activity is associated with cognitive processes that require externally directed attention, including working memory, relational integration, and task-set switching (Seeley et al., 2007). According to several authors, the default and control networks can exhibit an antagonistic relation at rest and during many cognitive tasks (Beaty et al., 2016). For example, Anticevic et al. (2012) note that during working memory tasks, the control network shows increased activity while the default network is deactivated. However, other researchers (Cristoff et al., 2009; Cocchi et al., 2013) have shown that the two networks cooperate in the execution of some cognitive tasks. Beaty et al. (2014) note that the default mode network and the executive control network cooperate in divergent thinking.

According to Beaty et al. (2016), creative thought involves dynamic interactions of large-scale brain systems, with the most compelling finding being that the default and executive control networks, which can show an antagonistic relation, tend to cooperate during creative cognition and artistic performance. Hence, brain systems are not always antagonistic and tend to cooperate during cognitive control (Cocchi et al., 2013). It has been proven that the default and control networks cooperate to generate and maintain an internal train of thought (Smallwood et al., 2012). This indicates that the ability to generate and sustain an internal train of thought unrelated to external reality frees an agent from the constraints of only acting on immediate, environmentally triggered events (ibid.). Smallwood et al. (2012) note that thought is produced through cooperation between autobiographical information provided by the default mode network and a frontal–parietal control network that helps sustain and buffer internal trains of thought against disruption by the external world. This cooperation suggests that creativity as generated in the brain is a combination of controlled and less controlled processes (Kühn, Ritter, and Muller et al., 2013).

3 IMPLICATIONS FOR ORGANIZATIONS

What could an understanding of the neural basis of creativity bring to organizations and managers? How can such knowledge be used to improve creativity and innovation in organizations? I address these questions by exploring how organizations can use neuroscience to advance creativity, innovation, and change, and how they can train employees to be more creative.

3.1 Using Neuroscience to Improve Innovation in Organizations

An interesting question is whether knowledge of the neural foundations of creativity could help to foster innovation in organizations. To the extent that creativity is the seed of innovation (Amabile, 1996) and organizations rely on the development and commercialization of new products to prosper and build a competitive advantage, such a question is more than relevant. Largely considered as a driver of innovation, growth, and societal development, creativity is often seen as "a vital means for organizations to thrive in dynamic environments, respond to unforeseen challenges, and proactively develop new capabilities" (Zhou and Hoever, 2014, p. 334). Scholars have studied the neural basis of processes, such as design and design thinking that could provide insights for innovation in the workplace. Design is an important factor that leads to change and innovation not only organizations but also in societies. Designing is a goal-directed, iterative, and creative activity that requires sustained cultivation of sophisticated cognitive competencies (Simon, 1977) and design thinking consists of approaching management problems in the same way that designers approach design problems (Dunne and Martin,

2006). This can particularly be used in the design of new products or in solving difficult organizational or societal problems.

For example, Alexiou et al. (2009) explored the neural basis of design thinking using fMRI. Their findings suggest that design and problem solving involve distinct cognitive functions associated with distinct brain networks. The authors show that Brodmann area 9 (BA9) on the right hemisphere is more activated in the design phase than in the problem-solving phase. There is a specialized network that includes the dorso-lateral prefrontal cortex, the dorsal areas in the anterior cingulate cortex, as well as areas in the medial temporal lobe and medial frontal gyrus, that is activated in design tasks. Hence, one may conclude that there is a neural basis to design thinking.

Luo et al. (2013) conducted two experiments in which college students were required to resolve new scientific innovation (NSI) problems (to which they did not know the answers) and old scientific innovation (OSI) problems (to which they did). Using what is called prototype heuristics, the authors suggest that innovation may engage automatic activation of a prototype such as a biological system to form novel associations between a prototype's function and problem solving. In Experiment 1, they found that the lingual gyrus (LG; BA18) might be related to prototype heuristics in college students resolving NSI after learning a relative prototype. In Experiment 2, the LG (BA18) and precuneus (BA31) were significantly activated for NSI compared to OSI when college students learned all prototypes one day before the test. These findings suggest that to some extent people can be trained to approach problems in more creative and innovative ways. After all, creative innovation implies that one discovers something entirely new or finds new combinations within familiar things.

3.2 Neuroscience of Creativity and Corporate Entrepreneurship

Corporate entrepreneurship includes the sets of activities firms undertake to stimulate innovation and risk taking to build competitive advantage (Zahra, Filatotchev, and Wright, 2009). It refers to "the process whereby the firms engage in diversification through internal development. Such diversification requires new resource combinations to extend the firm's activities in areas unrelated, or marginally related, to its current domain of competence" (Burgelman, 1983, p. 1349). The neuroscience of cre-ativity can provide insight into entrepreneurial cognitions and thereby help organizations become more entrepreneurial. To some extent, an organization constitutes an opportunity structure (Burgelman, 1983).

Hence, "expansion of current business and diversification through internal development are the major ways in which the opportunity seeking behavior of organizational participants can exert itself" (Burgelman, 1983, p. 1353). To prosper and build a competitive advantage, organizations need to become more entrepreneurial.

Indeed, entrepreneurial decision and action are thus informed by an unusually high degree of both creativity and intuition (Tomasino, 2007). Recently, there has been an application of neuroscience to entrepreneurship. Knowledge from the neuroscience of creativity can provide insights to help organizations create environments that allow employees not only to take personal initiatives but also act as "entrepreneurs." Several authors have emphasized the role of neuroscience in opportunity recognition (Beugré, 2016). Entrepreneurial opportunities can be discovered by using prior experiences and knowledge, intuition, and by connecting the dots (Baron, 2006). This implicates System 1 and its corresponding brain structures. Entrepreneurial opportunities can also be created through improvisation, effectuation, combination, and reflection (Sarasvathy, 2001), which involves System 2.

The application of neuroscience to entrepreneurship is appealing and several authors have suggested its use to assess entrepreneurial attitudes and behaviors such as alertness, entrepreneurial cognitions, opportunity creation and discovery, and entrepreneurial intention (De Holan, 2014; Krueger and Welpe, 2014; Nicolaou and Shane, 2014; Beugré, 2016). For example, De Holan (2014) notes that many of the phenomena studied in entrepreneurship invoke the mind of the entrepreneur. Consequently, they can be assessed with the technologies used to understand the brain.

3.3 Neural Basis of Change

Creativity helps people not only to cope with changes (Runco, 2004) but also to introduce changes. This is particularly true for organizations where managers and employees have to introduce new processes, products, and technologies. Although change is important for organizations to advance, it is equally true that employees resist change. The question then is: why do people resist change? Are there neural underpinnings that could help explain employees' resistance to change? Do people resist change because the human brain is wired to do so? Can findings from neuroscience provide additional insights on why people resist change?

It is obvious that change involves situations of uncertainty where the likelihood of different outcomes is unknown. Such uncertainty may lead to anxiety, fear, and other emotional reactions. Hence, uncertainty inherent in organizational change may activate particular brain regions.

For example, Camerer et al. (2004) describe the amygdala as an internal "hypochondriac," which provides "quick and dirty" emotional signals in response to potential fears (p. 561). The amygdala is a brain structure implicated in many different kinds of phenomena, such as attitudes, stereotyping, person perception, and emotion (Ochsner and Lieberman, 2001). To the extent that change increases uncertainty and "fear of the unknown," it may activate the anterior insula. The prefrontal cortex is also involved in uncertainty and change. It is considered the seat of willpower – the ability to take the long-term perspective in evaluating risks and rewards (Morse, 2006).

Another brain region implicated in reactions to uncertainty is the insula. "The insula is a region that processes information from the nervous system about bodily states, such as physical pain, hunger, the pain of social exclusion, disgusting odors, and choking" (Camerer et al., 2004, p. 568). It is possible that the anterior insula is activated when people experience new situations. Because change increases uncertainty and "fear of the unknown" (Camerer et al., 2004), it may activate the anterior insula. The prefrontal cortex, the seat of willpower, could be involved in uncertainty and change – its activation could help employees anticipate change.

Encountering new stimuli leads to more cognitive effort to process and adapt to them, whereas encountering familiar situations requires less cognitive effort. Hence, the use of System 1 and System 2 could help to improve our understanding of change. Incremental change or change containing familiar elements may be less threatening to employees than radical change that dramatically breaks existing processes and procedures. However, experience and practice may lead a rather complex behavior to become automatic. For example, Camerer et al. (2004) suggest that "when good performance becomes automatic (in the form of procedural knowledge), it is typically hard to articulate, which means human capital of this sort is difficult to reproduce by teaching others" (p. 560). Baron (2006) argues that experienced, repeat entrepreneurs generally search for opportunities in areas or industries where they are already knowledgeable. Such a comparison helps individuals make decisions fast and effortlessly for familiar situations, whereas novel ones require more cognitive effort. People may also tend to resist change because it involves the potential for loss, and according to prospect theory, they tend to be more sensitive to losses than gains (Kahneman and Tversky, 1979).

3.4 Creativity and Augmented Cognition

There is a growing interest in the use of neuroenhancement techniques to boost cognitive functions (Partridge et al., 2011). Augmented cognition can be described as efforts to boost the activity of neural structures that are responsible for creativity. It consists of interventions that make normal, healthy brains work better. For example, familiar substances such as coffee, tea, coca leaves, and alcohol can be used to alter brain chemistry and boost cognition (Farah, 2012, p. 579). Jarosz et al. (2012) tested the effects of moderate alcohol intoxication on a common creative problem-solving task – the Remote Associates Test (RAT). Individuals were brought to a blood alcohol content of approximately 0.075, and, after reaching peak intoxication, completed a battery of RAT items. Intoxicated individuals solved more RAT items, in less time, and were more likely to perceive their solutions as the result of a sudden insight. These findings pose the question of the link between substance abuse and creative cognition.

Neuroscience techniques, such as transcranial direct current stimulation (tDCS) are used to stimulate brain structures to enhance cognition. For example, Metuki, Sela, and Levador (2012) used tDCS on a sample of 21 participants and found that brain stimulation significantly enhanced solution recognition for difficult problems. Left dorsolateral prefrontal cortex executive control modulates semantic processing of verbal insight problems. Superior executive functioning may be detrimental to creative problem solving. Hence, in organizations, too many policies, guidelines, and rules regulating daily behavior can impede employee creativity. Chi and Snyder (2012) show that stimulating the brain using tDCS facilitates the resolution of apparently difficult problems. Creativity has also been associated with the loosening of some brain structures. Yaniv (2011) notes that "subtle frontally disinhibited behavior is conducive to creativity by allowing an uninterrupted flow of creative/spontaneous thought processing and opening up new avenues to problem solving" (p. 55).

Partridge et al. (2011) reviewed the media coverage of neuroenhancement drugs and found that 87 percent of media articles mentioned the prevalence of neuroenhancement, and 94 percent portrayed it as common, increasing or both, 66 percent referred to the academic literature to support these claims and 44 percent either named an author or a journal, 95 percent of articles mentioned at least one possible benefit of using prescription drugs for neuroenhancement, but only 58 percent mentioned any risks or side-effects. The authors found that most newspaper articles portray neuroenhancement as common or increasing in prevalence and focus more on its benefits than its potential side-effects.

Augmented cognition has applications in the US military DARPA (Defense Advanced Research Projects Agency). The DARPA Augmented Cognition program develops technologies that intend to transform the interactions between humans and machines. These technologies tend to adjust information systems to the capabilities and limitations of war-fighters. They are endowed with non-invasive sensors that are able to provide objective measures of neurophysiological responses of the users. Hence, they could facilitate the adaptation of the machines to users. Such augmented cognition technologies could be used in organizations to improve decision making and other organizational behaviors. This is particularly true in an era of Big Data and analytics. To analyze and make sense of the vast amounts of data organizations have at their disposal, they need managers and data professionals with strong cognitive abilities.

3.5 Neuroscience and Creativity Training

An interesting question is whether knowledge from neuroscience could be used to foster training in the generation of creative ideas. Onarheim and Friis-Olivarius (2013) collected empirical data and used their experiences from the Applied NeuroCreativity (ANC) program, taught in business schools in Denmark and Canada. Participants in the ANC program were first introduced to cognitive concepts of creativity, before applying these concepts to a relevant real-world creative problem. The novelty of the ANC program is that the conceptualization of creativity is built on neuroscience, and a crucial aspect of the course is giving the students a thorough understanding of the neuroscience of creativity.

The training focused on four main areas: (1) understanding creativity through the neuroscience of creativity; (2) understanding the difference between divergent and convergent thinking, and how the combination of these two is the source of creativity; (3) learning various creative tools, understanding why and how such tools work from a neurological perspective, and when to use them; and (4) applying the creative tools in practice, while reflecting and (if necessary) acting on the ongoing neurological processes, if these were limiting the creative process. Onarheim and Friis-Olivarius (2013) showed that ANC students gained more fluency in divergent thinking (a traditional measure of trait creativity) than those in highly similar courses without the neuroscience component, suggesting that principles from neuroscience can contribute effectively to creativity training and produce measurable results on creativity tests. The authors noted that the evidence presented indicated that the inclusion of neuroscience principles in a creativity course could in eight weeks increase divergent thinking skills with an individual relative average of 28.5 percent. They concluded that neuroscience-based creativity training works.

This is consistent with previous findings on the benefits of creativity training. Scott, Leritz, and Mumford (2004) note that creativity training provides some benefits to individual participants. Most successful programs were likely to focus on development of cognitive skills and the heuristics involved in skill application, using realistic exercises appropriate to the domain at hand. Sun et al. (2016) found that both the originality and the fluency of divergent thinking were significantly improved by training. Furthermore, functional changes induced by training were observed in the dorsal anterior cingulate cortex (dACC), the dorsolateral prefrontal cortex (dlPFC), and the posterior brain regions. They also found that the gray matter volume (GMV) was significantly increased in the dACC after divergent thinking training. These results suggest that the enhancement of creativity may rely not only on the posterior brain regions that are related to the fundamental cognitive processes of creativity but also on areas that are involved in top-down cognitive control, such as the dACC and the dlPFC.

Fink et al. (2010) investigated whether creative cognition could be enhanced through idea sharing and how performance improvements were reflected in brain activity. They had 31 participants generate alternative uses of everyday objects during fMRI recording. Participants also performed this task after a time period in which they had to reflect on their ideas or in which they were confronted with stimulus-related ideas of others. The authors found that cognitive stimulation was effective in improving originality, and this performance improvement was associated with activation increases in a neural network including right-hemispheric temporoparietal, medial frontal, and posterior cingulate cortices, bilaterally.

These findings provide evidence on improving the "creative potential" of the brain through training. If people can be trained to be creative using neuroscientific methods, then it is possible to introduce such techniques in R&D units within organizations. For example, idea generation sessions could be organized to stimulate both the default mode network and the executive control network. Could it be possible to stimulate employees' brains using transcranial magnetic stimulation methods to make them more creative? Although this could raise ethical issues, it could also provide some advantages in terms of the generation of novel and useful ideas. Creativity is enhanced by abstract representation of problem elements (Liberman and Trope, 1998). However, it is undermined by the provision of extrinsic rewards and the expectation of social evaluation

(Amabile, 1996). Therefore, managers should be cautious in administering rewards designed to acknowledge employees' creative ideas.

Although this chapter focuses on the neural basis of creativity, it is worth mentioning that several authors criticized any tendency to reduce the creative process to brain activation. Instead, they emphasize the importance of the social context and thereby advocate a systemic perspective. For example, Hennessey and Amabile (2010) advocate a system view of creativity and argue that "neurological events in the brain, behavioral manifestations of mental illness, or individual differences in personality must be studied not in isolation but in conjunction with the particular environment in which an individual's physical, intellectual, and social development has taken place" (p. 589).

Sawyer (2011) notes that "creativity is not dependent on any particular mental process or brain region" (p. 151). Despite the importance of research on the neural basis of creativity, we must acknowledge that the creative process cannot be limited to the activation of a specific brain structure. Rather, it is the result of a multitude of cognitive and neural processes. Hence, it could be better to reason in terms of neural networks of creative thinking rather than specific brain structures. In fact, creativity is a complex human ability and as such cannot be reduced to the activation of one single neuron. As Dietrich and Kanso (2010, p. 822) contend, "creative thinking does not appear to critically depend on any single mental process or brain region, and it is not especially associated with right brains, defocused attention, low arousal, or alpha synchronization, as sometimes hypothesized."

5. The neural basis of motivation and rewards

Psychologists, economists, organizational scholars, and philosophers have long debated the reason people engage in specific actions. One of the key elements of organizational life is why people engage in specific activities. Motivation is not a unitary concept and refers to the "orienting and invigorating impact, on both behavior and cognition, of prospective rewards" (Botvinick and Braver, 2015, p. 86). Most organizational scholars have acknowledged that motivating the workforce is arguably the number one problem facing many organizations today. What drives their behavior and what do they expect to gain by pursuing some actions and avoiding others? In other words, what motivates people to work toward certain goals and not others? Could people's desire to act in certain ways have neural underpinnings? This chapter addresses such questions.

1 UNDERSTANDING MOTIVATION

Motivation is often considered as the energy that drives someone to engage in a particular course of action. It refers to the forces within a person that affect the direction, intensity, and persistence of a voluntary behavior. Motivation concerns energy, direction, persistence, and equifinality – all aspects of activation and intention (White, 1959; Ryan and Deci, 2000; Latham and Pinder, 2005). It has been "a central and perennial issue in the field of psychology, for it is at the core of biological, cognitive, and social regulation" (Ryan and Deci, 2000, p. 68). Motivation is a universal characteristic of any living organism (Maslow, 1954). It is one of the widely studied concepts by psychologists and organizational behavior scholars and is valued because of its positive consequences.

In the workplace, motivation is defined as "a set of energetic forces that originate both within as well as beyond an individual's being, to initiate work-related behavior and to determine its form, direction, intensity, and duration" (Pinder, 1998, p. 11). Motivation cannot be observed; it can only be inferred from the person's actual behavior. It often leads to higher productivity and performance and lower turnover. Although the concept of motivation has been widely studied, the focus on its neural foundations is relatively new. To emphasize the importance of exploring the neural basis of motivation, scholars such as Berridge (2004) argue that "motivational concepts are needed to understand the brain just as brain concepts are needed to understand motivation" (p. 205). Berridge (2004) further argues that "trying to explain how the brain controls motivated behavior without motivational concepts is like trying to understand what your computer does without concepts of software" (ibid.). "Motivation occupies a center stage in the psychology and behavioral neuroscience of decision making" (Niv, Joel, and Dayan, 2006, p. 375). Neuroscientific research on motivation and control lies at the confluence of cognitive control and reward-based decision making.

"The anterior cingulate cortex is responsible for learning and selecting high-level behavioral plans that provide the meaning behind, and thus the motivation for, our moment-to-moment actions" (Holroyd and Yeung, 2012, p. 128). Using functional magnetic resonance imaging (fMRI), LeBouc and Pessiglione (2013) scanned human participants while they made a physical effort in a collaborative or competitive context. They found that motivation was primarily driven by personal utility, which was reflected in brain regions devoted to reward processing (the ventral basal ganglia). In their study, participants who departed from utility maximization and worked more in collaborative situations showed greater functional activation and anatomical volume in a brain region implicated previously in social cognition (the temporoparietal junction).

The following sections explore the neural basis of need theories and process theories of motivation. Separating motivation theories into need theories and process theories follows a long tradition in organizational behavior research. In need theories, one of the key arguments is that motivation is a process by which people satisfy unfulfilled needs. What drives a person's behavior is the desire to satisfy a need. Process theories, however, consider motivation as the product of cognitive processing and reasoning. People engage in particular behaviors after having cognitively processed some stimuli they deem important.

2 NEURAL BASIS OF MOTIVATION THEORIES

There are diverse theories of motivation attempting to explain why people behave the way they do, particularly in organizations. This diversity of motivation theories led Wood et al. (2015, p. 145) to conclude that "no single motivational theory exists." The analysis of the neural basis of motivation in organizations could benefit from evolutionary neuroscience. According to Wood et al. (2015, p. 153), "evolutionary neuroscience suggests that our motivation comes from two phylogenetic sources that can be summarized as self-interest and other self-interest." This section explores the neural basis of two sets of motivation theories applied to organizations: need theories and process theories.

2.1 Overview of Need Theories of Motivation

This section explores the neural basis of need theories of motivation. Prior to doing so, the section briefly describes each of these theories, which are summarized in Table 5.1. Four theories (Maslow's hierarchy of needs, Alderfer's ERG theory, McClelland's need theory, and Lawrence and Nohria's need theory) are briefly discussed. The common thread of need theories is that people are motivated to satisfy particular needs. In other words, the desire to satisfy specific needs is the driver of human behavior. A need is often defined as a deficiency – something we lack and for which we will take an action. The reasoning is that a need creates a tension and the person is motivated to reduce the tension by fulfilling the need.

2.1.1 Maslow's hierarchy of needs

Maslow's theory is the most popular need theory. Maslow identified five types of needs – physiological needs, safety needs, social needs, esteem needs, and self-actualization needs – which form a hierarchy (see Table 5.1). When a need is not satisfied, the person experiences a tension and acts to fulfill this need. Once the need is satisfied, it no longer motivates the person. Since Maslow considers needs as forming a hierarchy, people have to satisfy lower-level physiological needs before satisfying higher-level needs. As the person satisfies that need, the next higher need in the hierarchy becomes the primary motivator. According to Maslow, self-actualization represents growth of an individual toward the fulfillment of the highest needs that give meaning to life. It is considered the highest-order need in Maslow's hierarchy. When people satisfy their self-actualization need, they are more inclined to seek situations that provide the opportunity to satisfy this need, unlike the satisfaction of lower-level needs, such as physiological or safety needs. Hence, the satisfaction-progression hypothesis does not apply to the satisfaction of self-actualization needs.

2.1.2   Alderfer's existence, relatedness, and growth need theory

Alderfer (1969, 1972) developed a three-factor need theory involving existence, relatedness, and growth needs (ERG) (see Table 5.1). Existence needs include all physiological and material needs. They correspond to physiological and safety needs in Maslow's theory of need. Relatedness needs include the need to establish relationships with others and correspond to social needs in Maslow's hierarchy of needs. Growth needs correspond to needs that help a person utilize his or her full potential and capabilities, and to Maslow's esteem needs. ERG theory states that a person can be motivated by several needs at the same time. For instance, an employee may try to satisfy his or her growth need without having completely satisfied his or her relatedness need. Unlike Maslow's theory, ERG theory includes a frustration-regression hypothesis. According to this hypothesis, when an individual is unable to satisfy a higher-level need, he or she becomes frustrated and regresses to the next lower-level need. For instance, if existence and relatedness needs have been satisfied but the growth need has not, the individual will become frustrated and relatedness needs will again emerge as the dominant source of motivation.

2.1.3   McClelland's need theory

McClelland identified three types of need that drive human behavior: the need for achievement, need for affiliation, and need for power (see Table 5.1). According to McClelland, these needs are learned through childhood learning, parental influences, and social norms. McClelland (1961, 1965) considered the need for achievement (corresponding to Maslow's need for self-actualization) as one of the main drivers of entrepreneurship and prosperity of societies and defined it as the desire to attain an inner feeling of personal accomplishment, which is satisfied primarily by an intrinsic sense of success and excellence rather than extrinsic rewards. People with a high need for achievement prefer challenging assignments, competition, and feedback on their actions. They are mostly motivated by the expectations of satisfying their need for achievement, whereas money tends to motivate people with a low need for achievement.

The need for affiliation refers to the desire to seek approval and acceptance from others. People with a strong need for affiliation want to have positive relationships with others. The need for affiliation is similar to Maslow's social needs and Alderfer's relatedness needs. The third need, the need for power, refers to the willingness to control one's environment. People with a high need for power will seek positions of power and prefer to stay in control. McClelland identified two types of power: personalized power and socialized power. People with personalized power use power to advance their own interests, whereas people with socialized power use power to advance the group or the organization's interests.

2.1.4 Lawrence and Nohria's theory of drive

In their seminal book, *Driven: How Human Nature Shapes Our Choices*, Lawrence and Nohria (2002) identified four main factors that drive human behavior: (1) the drive to acquire; (2) the drive to defend; (3) the drive to bond; and (4) the drive to comprehend (or learn). The drive to acquire refers to the natural tendency human beings have to acquire goods that improve their well-being. For example, working can be a means for people to acquire the goods they value. According to Lawrence and Nohria (2002), the drive to acquire is relative and people tend not to have enough of the goods they value. The drive to defend is rooted in the basic fight-or-flight response (Lawrence and Nohria, 2002; Nohria, Groysberg, and Lee, 2008). The drive to acquire and the drive to defend could be related to existence needs. For example, people strive to acquire the resources that will help them satisfy their basic needs. The drive to defend could refer to safety needs that help people protect themselves against events that could physically or psychologically harm them.

The drive to bond is similar to social needs in Maslow's hierarchy of needs, relatedness needs in Alderfer's three-factor theory, or the need for affiliation in McClelland's three-factor needs theory. Human beings are social animals and strive to be included in social groups and connect with others. The drive to comprehend pushes humans to collect information, assess the needs of a situation, examine their environment, and make observations about explanatory ideas and theories to appease curiosity and make sound judgments (Lawrence and Nohria, 2002). It is similar to Maslow's self-actualization needs and Alderfer's growth needs. McClelland's needs for power and achievement could be included in the drive to comprehend. According to Lawrence and Nohria (2002) and Nohria et al. (2008), each of the four drives is independent and cannot be ordered hierarchically or substituted one for the other.

Nohria et al. (2008) note that motivation is created in the brain when dopamine is released and takes a specific direction toward the mesolimbic pathway and then spreads to other areas in the brain like the cerebral cortex. Dopamine is the reward and punishment transmitter and plays a critical role in positive motivation and aversive motivation. Drive and motivation are central to affective neuroscience (Kringelbach and Berridge, 2016).

Table 5.1   Comparison of need theories

Maslow	Alderfer	McClelland	Lawrence and Nohria
Physiological	Existence		Drive to acquire
Safety			Drive to defend
Social	Relatedness	Affiliation	Drive to bond
Esteem	Growth		
Self-actualization		Need for power	
		Need for achievement	Drive to learn

2.2   Neural Basis of Need Theories of Motivation

Although need theories of motivation have been extensively studied and figure prominently in organizational behavior textbooks, there are few investigations on whether these theories could be explained through the lens of neuroscience. However, there are some findings on the neural basis of reward, valuation, and reinforcement that could provide insights into the neural basis of need theories of motivation. For example, Kim, Reeve, and Bong (2016) identified three brain circuits involved in human motivation: (1) the reward circuit; (2) the value-based decision making pathway; and (3) the self-regulation/self-control network.

Takeuchi et al. (2014) used the neuroimaging analysis technique voxel-based morphometry and a questionnaire (achievement motivation scale) to measure individual achievement motivation and investigated the association between regional gray matter density (rGMD), self-fulfillment achievement motivation (SFAM), and competitive achievement motivation (CAM) across the brain in healthy young adults (age 21.0 ± 1.8 years, men [$n = 94$], women [$n = 91$]). SFAM and rGMD significantly and negatively correlated in the orbitofrontal cortex (OFC). CAM and rGMD significantly and positively correlated in the right putamen, insula, and precuneus. These results suggest that the brain areas that play central roles in externally modulated motivation (OFC and putamen) also contribute to SFAM and CAM, respectively, but in different ways. Furthermore, the brain areas in which rGMD correlated with CAM are related to cognitive processes associated with distressing emotions and social cognition, and these cognitive processes may characterize CAM.

Quirin et al. (2013) used fMRI to identify neural structures involved in power versus affiliation motivation based on an individual differences approach. Seventeen participants provided self-reports of power and affiliation motives and were presented with love, power-related, and control movie clips. The power motive predicted activity in four clusters within the left prefrontal cortex while participants viewed power-related film clips. The affiliation motive predicted activity in the right putamen/pallidum while participants viewed love stories. Power-related versus affiliation-related social motivations had differential brain networks.

People have a natural tendency to engage in actions that benefit others. The desire to engage in such actions is termed prosocial motivation. "Prosocial motivation is the desire and drive to benefit others, prosocial behaviors are the acts that benefit others, and prosocial impact is the awareness that one's actions have succeeded in benefiting others" (Bolino and Grant, 2016, p. 602). Prosocial motivation refers to the desire to benefit others or expend effort out of concern for others (Grant, 2008). Research shows that cooperation engages several areas in the brain's reward circuitry including the nucleus accumbens, the caudate nucleus, and the ventromedial prefrontal cortex (see Chapter 8). Tabibnia and Lieberman (2007) note that fairness and cooperation are rewarding in themselves because they activate the brain's reward circuitry. Breitner et al. (2001) found that brain regions that respond to receiving money are the same as those that react to attractive faces, chocolate, cocaine or morphine, music, revenge, sex, and sports cars.

Izuma, Saito, and Sadato (2008) studied the neural basis of monetary and social reward. They had 19 subjects participate in fMRI experiments and found that acquisition of one's good reputation activated the striatum, and this overlapped with the areas activated by monetary reward. These findings indicate that the acquisition of reputation motivates the same neural circuitry as receiving monetary reward. The existence of a "common neural currency" implies that there is a neural circuitry responsible for reward, be it monetary or social.

Individuals are often motivated to engage in actions intended to benefit others (Bolino and Grant, 2016, p. 599). Three areas of research on prosocial motivation in organizations are: (1) prosocial motives (the desire to benefit others or expend effort out of concern for others); (2) prosocial behaviors (acts that promote/protect the welfare of individuals, groups, or organizations); and (3) prosocial impact (the experience of making a positive difference in the lives of others through one's work). Even in the absence of social pressure, individuals routinely forego personal gain to share resources with others. Prosocial behavior may reflect an intrinsic value placed on social ideals such as equity and charity.

Zaki and Mitchell (2011) found that making equitable interpersonal decisions engaged neural structures involved in computing subjective value, even when doing so required forgoing material resources. They found that acting prosocially engaged the orbitofrontal cortex, whereas engagement of the anterior insula accompanied inequitable decision making. By contrast, making inequitable decisions produced activity in the anterior insula, a region linked to the experience of subjective disutility. Moreover, inequity-related insula response predicted individuals' unwillingness to make inequitable choices. Social principles have an intrinsic value in themselves. Altruistic motivation has neural foundations (Mathur et al., 2010) and may lead people to cooperate with and help others in organizations. This is particularly important because "humans have evolved to live cooperatively in social groups in which people take care of each other" (Crocker, Canevello, and Brown, 2017, p. 315).

2.3 Overview of Process Theories of Motivation

Process theories of motivation contend that behavior is a function of beliefs, expectations, and values. Behavior is viewed as a result of rational and conscious choices. This section presents a brief overview of four process theories of motivation: goal-setting theory, equity theory, valence/expectancy theory, and self-determination theory.

2.3.1 Goal-setting theory

Locke (1996) and Locke and Latham (1990, 2002) developed goal-setting theory to explain the importance of goals as the motivator of human behavior. In goal-setting theory, a goal is defined as what an individual is trying to accomplish; it is the objective or aim of an action. "Goals are a key element in self-regulation" (Locke and Latham, 2006, p. 265). According to Locke, goal setting has four motivational mechanisms: (1) goals direct actions by focusing a person's attention on what is relevant; (2) goals regulate effort and motivate a person to act; (3) goals increase persistence, which represents the effort expended on a task over an extended period of time – persistent individuals tend to view obstacles as challenges to be overcome rather than reasons to give up; (4) goals foster strategies and action plans – setting goals allows people to develop concrete actions to attain them.

Locke (1996) also notes that the conditions that enhance the motivational power of a goal include difficulty of the goal, specificity of the goal, opportunities for feedback, goal acceptance, and goal commitment. Managers may use goal-setting theory as a motivational technique. Research on goal-setting theory showed that specific, difficult goals lead to a higher level of task performance than do easy goals or vague, abstract goals such as the exhortation to "do one's best" (Locke and Latham, 1990, 2002; Locke, 1996). Goal-setting is an open theory and has been applied to several areas including sports, education, personal lives as well as in organizations.

2.3.2 Equity theory

Adams (1965) developed equity theory, which is a theory of social comparison. The theory postulates that the individual compares himself or herself to another individual. In so doing, the person compares his or her input/output ratio to that of another person – the comparison other or the referent. Three elements – input, output, and referent – are important in equity theory. Input is what the person brings to the exchange relationship. Examples of input include effort, education, and experience. Output is what the person gets from the exchange relationship. Examples of output include compensation, bonus, and promotion. The referent is the comparison other – the individual that the person compares himself or herself to.

An individual experiences feelings of equity when the two ratios are equal. However, when the person's ratio is greater than that of the comparison other, he or she experiences feelings of advantageous inequity. According to Adams (1965), the possible reaction to feelings of advantageous inequity is the sense of guilt. However, in reality, people tend to rationalize advantageous inequity. If the individual's ratio is less than that of the comparison other, he or she experiences feelings of disadvantageous inequity. Disadvantageous inequity is less tolerable than positive inequity and can be reduced in a variety of ways: reducing one's inputs, increasing one's inputs, increasing one's output, decreasing one's output, psychologically distorting inputs and outputs of the comparison other, quitting the exchange relationship, or changing the comparison other. In equity theory, the driver of human behavior is the feeling of equity. People are motivated when they experience a sense of equity and demotivated when they suffer inequity.

2.3.3 Valence/expectancy theory

The main assumption of expectancy theory is the rather simple concept that an individual's behavior is a function of the degree to which the behavior is instrumental in the attainment of some outcomes, and the evaluation of these outcomes (Vroom, 1964). In developing this theory, Vroom (1964) defined motivation as the force impelling a person to perform a particular action, as determined by the interaction of (1) the person's expectancy that his or her act will be followed by a particular outcome; and (2) the valence of that (first-level) outcome. There are three components of expectancy theory: expectancy, instrumentality, and valence. Expectancy refers to the relationship between effort and performance. For instance, an employee may clearly see that making an effort in a given task will lead to a certain level of performance. Instrumentality refers to the relationship between performance and rewards. Valence refers to the value of the reward.

Expectancy theory proposes a causal relationship between expectancy attitudes and motivation (Vroom, 1964; Lawler, 1973). Research on expectancy theory has shown that expectancy attitudes and job performance are positively related (Vroom, 1964; Lawler, 1973), whereas other studies did not find a positive relationship between expectancy attitudes and job performance (Reinharth and Wahba, 1975). To some extent, managers have implemented expectancy theory by clearly explaining the link between effort and performance, linking performance to rewards (performance-based compensation), and making rewards attractive to employees. In addition to organizations, expectancy theory has been applied to several domains including education, sales force motivation, and even alcohol consumption.

2.3.4 Self-determination theory

Self-determination theory is an approach that explores human motivation (Ryan and Deci, 2000). It identifies three basic needs: competence, relatedness, and autonomy. Factors that increase these three needs augment intrinsic motivation, whereas factors that decrease them diminishes intrinsic motivation. The theory essentially contends that "the fullest representations of humanity show people to be curious, vital, and self-motivated. At their best, they are agentic and inspired, striving to learn; extend themselves; master new skills; and apply their talents responsibly" (Ryan and Deci, 2000, p. 68). According to self-determination theory, autonomy is a basic psychological need that plays a key role in intrinsic motivation (Meng and Ma, 2015). An extension of self-determination theory is cognitive evaluation theory, which contends that two psychological needs underlie intrinsic motivation: (1) the need for self-determination; and (2) the need for competence (Deci and Ryan, 1985).

2.4 Neural Basis of Cognitive Theories of Motivation

The common thread of process theories of motivation is the assumption that motivation is a cognitive process. Hence, these theories could be studied from a neuroscientific perspective to the extent that cognitive processes are substantiated by neural substrates (Lieberman, 2007a,

2007b). For example, research on the neural basis of rewards could provide insights into the neural foundations of expectancy theory. Rewards refer to stimuli that positively reinforce the frequency or intensity of a behavior pattern. Primary rewards include food, water, and sexual stimuli. Secondary rewards include cultural factors and money and acquire their secondary value by being associated with primary rewards. Secondary rewards reinforce behavior after they have been learned. Research in neuroscience shows that the brain regions involved in reward include the ventral striatum, the dorsolateral prefrontal cortex, the orbital prefrontal cortex, the nucleus accumbens, and the amygdala (Schultz, 2002; Camerer et al., 2004; McClure, York, and Montague, 2004; Walter et al., 2005; Delerck, Boone, and Emonds, 2013). Research in neuro-science has indicated that neural activity in the striatum has consistently been shown to scale with the value of anticipated rewards (Miller et al., 2014). Miller et al. (2014) noted that responses in the caudate and putamen increased with motivation, whereas nucleus accumbens activity increased with expected reward.

Murayama et al. (2015) examined the neural correlates of the facilita-tive effects of self-determined choice using fMRI. Participants played a game-like task involving a stop-watch with either a stop-watch they selected (self-determined choice condition) or one they were assigned without choice (forced choice condition). They found that self-determined choices enhanced performance on the stop-watch task, even though the choices were clearly irrelevant to task difficulty. Neuro-imaging results showed that failure feedback, compared with success feedback, elicited a drop in the ventromedial prefrontal cortex (vmPFC) activation in the forced choice condition, but not in the self-determined choice condition, indicating that negative reward value associated with the failure feedback vanished in the self-determined choice condition. Moreover, the vmPFC resilience to failure in the self-determined choice condition was significantly correlated with increased performance. Striatal responses to failure and success feedback were not modulated by the choice condition, indicating the dissociation between the vmPFC and striatal activation pattern. These findings suggest that the vmPFC plays a unique and critical role in the facilitative effects of self-determined choice on performance.

Neuroeconomic evidence also indicates that people develop a propen-sity to experience direct discomfort when they spend money (Camerer et al., 2004; McClure et al., 2004). When participants earned money by responding correctly to a stimulus rather than just receiving equivalent rewards with no effort, there was a greater activity in the striatum, a reward region of the brain (Camerer et al., 2004; McClure et al., 2004).

Mizuno et al. (2008) used fMRI to study the motivation to learn and the motivation to gain monetary rewards. They found that the motivation to learn correlates with bilateral activity in the putamen, and the higher the reported motivation, the greater the change in blood oxygenation level-dependent (BOLD) signals within the putamen. Monetary motivation also activated the putamen bilaterally, though the intensity of activity was not related to the monetary reward, indicating that earned money is more rewarding in the brain than unearned money (Camerer et al., 2004).

These findings may have direct implications for managers. Tying compensation to performance may be more rewarding for employees than compensation that is not performance based. People respond to monetary incentives and consequently organizations have adopted performance-based compensation systems to motivate employees. The rationale is that employees will be better motivated and perform better when their compensation is tied to their actual performance. However, we must acknowledge that the relationship between money, motivation, and work performance is more complicated than originally thought. People respond to monetary incentives because the limbic system quickly becomes accustomed to new stimuli and reacts only to the unexpected, such as a financial windfall (Camerer et al., 2004; McClure et al., 2004). Un-expected rewards or rewards higher than expected produce a phasic increase in the firing rate of the dopamine neurons at the time of their delivery (Caldu and Dreher, 2007). Haruno and Kawato (2006) conducted an event-related fMRI study and found that neural correlates of the stimulus–action–reward association reside in the putamen, whereas a correlation with reward prediction error was found largely in the caudate nucleus and ventral striatum.

In organizations, the prospect of a pay increase or a promotion may motivate employees. Employees are often motivated by what they see as gains and disappointed by what they see as losses. Dickaut et al. (2003) found more activity in the orbitofrontal cortex when subjects were thinking about gains compared to losses, and more activity in the inferior parietal and cerebellar areas when subjects were thinking about losses. When people experience pleasure or anticipate pleasure, the nucleus accumbens is activated. However, the insula is activated when people experience pain, taste something bad, anticipate pain or see a disgusting picture. Unexpected rewards tend to activate the striatum more than rewards that are anticipated. Employees may respond positively to bonuses and rewards that are unexpected because they activate the brain's reward circuitry. The "juice reward" experiment, where the promise of a valued reward activates the brain's reward circuitry, may apply to organizations.

Promising a pay increase, a bonus, or a promotion in exchange for good performance may motivate employees to work harder because the expectation of reward is likely to stimulate the nucleus accumbens, the seat of dopamine receptors. The expectation of reward is so powerful that people tend to downplay the potential risks when they see opportunities for large rewards. Daw and Shohami (2008) note that converging evidence suggests that midbrain dopamine neurons signal a reward prediction error, allowing an organism to predict and to act to increase the probability of reward in the future. Dopamine and the striatum play an important role in motivation and action. Dopamine is involved in reinforcement learning. The concept of "incentive motivation" implies doing something to receive an expected reward.

Robbins and Everitt (1996) note that nucleus accumbens dopamine activity is enhanced during appetitive phases of motivated behavior (p. 231). An appetitive phase includes behaviors of approaching or seeking a reward. Appetitive clues prompt sustained orienting responses that extend perceptual processing and facilitate action selection (Lang and Bradley, 2013). Appetitive motivation is considered as positive motivation (Salomone, 1994) and the brain structures involved in incentive motivation include the amygdala and the nucleus accumbens. The nucleus accumbens is also involved in aversive motivation. According to Bromberg-Martin, Matsumoto, and Hikosaka (2010), some dopamine neurons encode motivational value, supporting brain networks for seeking, evaluation, and value learning. Others encode motivational salience, supporting brain networks for orienting, cognition, and general motivation. We must acknowledge, however, that there are also non-monetary incentives that activate the brain's reward circuitry.

3   NEURAL BASIS OF EXTRINSIC AND INTRINSIC MOTIVATION

Most motivational theories focus on external rewards as the drivers of human behavior, particularly in work settings. People work to satisfy basic needs or to achieve desirable outcomes. However, some theories, such as self-determination theories and components of need theories, argue that people work because of the enjoyment experienced by performing certain tasks. Hence, Deci, Koestner, and Ryan (1999) focus their attention on the concepts of intrinsic and extrinsic rewards. The section addresses the neural underpinnings of these two types of motivation.

3.1 Understanding Intrinsic and Extrinsic Motivation

Harlow (1950) coined the term intrinsic motivation to describe his observation that primates would persist in playing with mechanical puzzles even in the absence of external rewards. He observed that the introduction of rewards for playing led these primates to decrease their spontaneous manipulative explorations, relative to those not exposed to external rewards. Intrinsic motivations are motivations based on mechanisms that drive learning of skills and knowledge, and the exploitation and energization of behaviors that facilitate this, on the basis of the levels and the variations of such skills and knowledge directly detected within the brain. An intrinsic motivation is the motivation to voluntarily engage in a task for the inherent pleasure and satisfaction derived from the task itself (Deci, 1971). Intrinsic motivation also refers to the inherent tendency to seek out novelty and challenges, to extend and exercise one's capacities, to explore, and to learn (Ryan and Deci, 2000, p. 70). People perform some tasks for pleasure and enjoyment. For example, tasks such as reading a comic book, writing prose, or volunteering for Sunday school could be performed for their own sake. Intrinsic motivation is a complex cognitive, affective, and behavioral phenomenon that is likely mediated by multiple neural structures and processes (Di Domenico and Ryan, 2017).

"Extrinsically motivated behaviors are governed by the prospect of instrumental gain and loss (e.g., incentives), whereas intrinsically motivated behaviors are engaged in for their very own sake (e.g., task enjoyment), not being instrumental toward some other outcome" (Cerasoli, Nicklin, and Ford, 2014, p. 980). Cerasoli et al. (2014) found that intrinsic motivation is a good predictor of performance. They also found that the link between intrinsic motivation and performance remained in place whether incentives were given to participants or not. However, intrinsic motivation was less important to performance when incentives were directly tied to performance and was more important when incentives were indirectly tied to performance. Considered simultaneously through meta-analytic regression, intrinsic motivation predicted more unique variance in quality of performance, whereas incentives were a better predictor of quantity of performance. With respect to performance, incentives and intrinsic motivation were not necessarily antagonistic and were best considered simultaneously.

3.2 The Undermining Effect of Extrinsic Rewards

Several authors have underscored the undermining effect or crowding out effect of extrinsic rewards. A theory that has addressed this issue is cognitive evaluation theory. According to cognitive evaluation theory, the introduction of an extrinsic reward for a job previously done for its own sake can reduce the motivation to perform it. Indeed, external rewards undermine intrinsic motivation (Deci et al., 1999). Murayama et al. (2010) found that performance-based monetary reward undermines intrinsic motivation, as assessed by the number of voluntary engagements in the task. They found that activity in the anterior striatum and the prefrontal areas decreased with this behavioral undermining effect. "Undermining effect" or "motivation crowding-out effect" is the tendency for external rewards to reduce the motivation to accomplish a task once done voluntarily. The undermining effect has implications for work motivation.

Several other authors have also confirmed the crowding-out effect of extrinsic motivation (Chamorro-Premuzic, 2013; Ma et al., 2014). In this regard, money may serve as a demotivator under certain circumstances (Chamorro-Premuzic, 2013). Hence, the source of motivation is in the brain. Ma et al. (2014) conducted an electrophysiological study with a simple but interesting stop-watch task to explore to what extent the performance-based monetary reward undermines individuals' intrinsic motivation toward the task. The electrophysiological data showed that the differentiated feedback-related negativity amplitude toward intrinsic success/failure divergence was prominently reduced once the extrinsic reward was imposed beforehand. However, such a difference was not observed in the control group, in which no extrinsic reward was provided.

Other research shows mixed results related to the undermining effect of extrinsic rewards. Marsden et al. (2015) used fMRI to examine the neural substrates of intrinsic motivation, operationalized as the free-choice time spent on a task when this was not required, and tested the neural and behavioral effects of external reward on intrinsic motivation. They found that increased duration of free-choice time was predicted by generally diminished neural responses in regions associated with cognitive and affective regulation. By comparison, the possibility of additional reward improved task accuracy, and specifically increased neural and behavioral responses following errors. Those individuals with the smallest neural responses associated with intrinsic motivation exhibited the greatest error-related neural enhancement under the external contingency of possible reward. Together, these data suggest that human performance is guided by a "tonic" and "phasic" relationship between the neural substrates of intrinsic motivation (tonic) and the impact of external incentives (phasic).

3.3 Neuroscience of Intrinsic and Extrinsic Motivations

Di Domenico and Ryan (2017) reviewed the literature on intrinsic motivation and found converging evidence suggesting that intrinsically motivated exploratory and mastery behaviors are phylogenetically ancient tendencies that are subserved by dopaminergic systems. They also found that several studies suggest that intrinsic motivation also is associated with patterns of activity across large-scale neural networks, namely those that support salience detection, attentional control, and self-referential cognition.

Albrecht et al. (2014) investigated the neural processes underlying the effects of monetary and verbal rewards on intrinsic motivation in a group of 64 subjects applying fMRI. They found that when participants received positive performance feedback, activation in the anterior striatum and midbrain was affected by the nature of the reward; compared to a non-rewarded control group, activation was higher while monetary rewards were administered. However, they did not find a decrease in activation after reward withdrawal. In contrast, the authors found an increase in activation for verbal rewards: after verbal rewards had been withdrawn, participants showed a higher activation in the aforementioned brain areas when they received success compared to failure feedback. They further found that while participants worked on the task, activation in the lateral prefrontal cortex was enhanced after the verbal rewards were administered and withdrawn. Jin, Yu, and Ma (2015) suggest that intrinsic motivation exists and can be detected at the neural level.

Using event-related fMRI, Lee and Reeve (2012) scanned 16 healthy human subjects while they imagined the enactment of volitional, agentic behavior on the same task but either for a self-determined and intrinsically motivated reason or for a non-self-determined and extrinsically motivated reason. Results showed that the anterior insular cortex (AIC), known to be related to the sense of agency, was more activated during self-determined behavior while the angular gyrus, known to be related to the sense of loss of agency, was more activated during non-self-determined behavior. Furthermore, AIC activities during self-determined behavior correlated highly with participants' self-reported intrinsic satisfactions.

W. Lee et al. (2012) scanned participants' neural activity when they decided to act for intrinsic reasons versus when they decided to act for extrinsic reasons. Intrinsic reasons for acting recruited insular cortex activity more, while extrinsic reasons for acting recruited posterior cingulate cortex activity more. The results demonstrate that engagement decisions based on intrinsic motivation are determined more by weighing the presence of spontaneous self-satisfaction such as interest and enjoyment, while engagement decisions based on extrinsic motivation are determined more by weighing socially acquired stored values as to whether the environmental incentive is attractive enough to warrant action.

Jin et al. (2015) employed event-related potentials to investigate the neural disparity between an interesting stop-watch (SW) task and a boring watch-stop task (WS) to understand the neural mechanisms of intrinsic motivation. Their data showed that, in the cue priming stage, the cue of the SW task elicited smaller N200 negativity amplitude (an event-related potential component) than that of the WS task. Furthermore, in the outcome feedback stage, the outcome of the SW task induced smaller feedback-related negativity amplitude and larger P300 positivity amplitude than that of the WS task. The authors showed that intrinsic motivation can be located at the neural level.

Organizations may combine both extrinsic and intrinsic rewards in motivating employees. From a neuroscience perspective, these two sets of motivations activate similar brain structures. For example, Lee (2016) notes that the study of the neural basis of extrinsic motivation shows that the striatum is central to extrinsic motivation but it is also activated in intrinsic motivation. Lee suggests that insular cortex activity, known to be related to intrinsic enjoyment and satisfaction, is a unique neural component of intrinsic motivation.

Exploring the neural foundations of motivation adds a third layer of analysis to the study of motivation (Figure 5.1). At the behavioral level, motivation encompasses choice, duration, effort, frequency, and regulation. At the psychological level, motivation deals with goal, value, pleasure, and aversion (pain). People are motivated to accomplish specific goals as determined by goal-setting theory. People also value certain outcomes compared to others. In this regard, they experience pleasure when they receive outcomes they value, and pain when they do not. People are also motivated to avoid outcomes they do not desire. The neural level of analysis includes the reward circuitry of the brain (amygdala, anterior cingulate cortex, dorsolateral prefrontal cortex, orbitofrontal cortex, and ventral striatum). Kim (2013) also observed this classification; however, he did not identify pain (aversive motivation) at the psychological level.

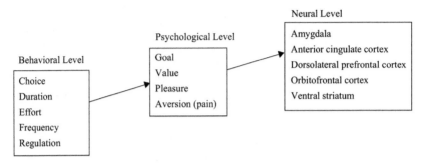

Figure 5.1   Levels of analysis of the neural basis of motivation

6. The neural basis of leadership

Leadership is one of the most exciting and researched topics in management, organizational behavior, and social psychology. The question of what makes some people display better leadership abilities than others has been the focus of attention in the literature. Likewise, the question of whether leaders are born or made has also been addressed in early research on leadership. Early theories of leadership have focused on the "big man," that is, on personality traits that differentiate leaders from non-leaders (Stogdill, 1948; Zaccaro, 2007). Although the trait approach to leadership has dominated early research paradigms, it has somehow faded and given rise to research on actual behaviors displayed by leaders. Zaccaro (2007) concludes that the trait approach has failed because of its "inability to offer clear distinctions between leaders and nonleaders and for its failure to account for situational variance in leadership behavior" (p. 6). Hence, behavioral and contingency theories of leadership dominated leadership research in the 1950s and 1960s (Fiedler, 1964; Hershey and Blanchard, 1969). Behavioral theories of leadership focused on actual behaviors displayed by leaders. Such behaviors include consideration and initiating structure (Fleishman, 1953b, 1957). A considerate leader emphasizes good relationships with followers, whereas a leader who focuses on initiating structure is concerned with roles, task requirements, and completion. Contingency theories tend to focus on the circumstances under which leadership occurs and how they affect the leader's behavior. For example, Fiedler's (1967) contingency model of leadership contends that leadership effectiveness depends on matching the leader's behavior to the situation. Further developments have witnessed the emergence of several new leadership theories, such as charismatic leadership, transactional leadership, transformational leadership, servant leadership, and authentic leadership, to list the most popular.

More recently, some authors (Waldman, Balthazard, and Peterson, 2011a, 2011b; Lee, Senior, and Butler, 2012a; Waldman and Balthazard, 2015) have called for the integration of a neuroscience perspective into the study of leadership. Such a perspective can be traced back to Mintzberg (1976), who was one of the first authors to allude to the role of the brain in explaining leadership in organizations. In a *Harvard Business Review* article, Mintzberg (1976) suggested that left brain/right brain differences may be relevant to management and leadership. He argued that some managers may be more inclined toward using one hemisphere rather than the other and this could affect the way they manage people and organizations. However, Mintzberg's observation did not gain traction in the 1980s and 1990s although scholars knew of the differences between the left and right hemispheres (Hellige, 1990). Hellige (1990) notes that the left hemisphere is largely responsible for the rational or analytical consideration of details when making decisions and holistic processing of information, whereas the right hemisphere is responsible for emotion-processing information.

A neuroscientific approach could help "both theory forming and theory honing because it will help to winnow theories about which cognitive processes facilitate various social behaviors" (Butler and Senior, 2007, p. 5). In addition, using neuroscience "may reveal insights into effective forms of leadership" (Balthazard et al., 2012, p. 255). It could also help leadership theory move ahead and avoid atrophy. This is particularly important because "organizational cognitive neuroscience is applied social cognitive neuroscience" (Lee et al., 2012a, p. 215). Neuroscience techniques can be combined with more traditional leadership develop- ment techniques, such as 360-degree feedback, to train leaders. Identify- ing a neurological profile, for example, could be a fruitful avenue of research.

The study of the neural foundations of leadership can be construed as an extension of the genetic studies on leadership (Johnson et al., 1998; Chaturvedi et al., 2011; De Neve et al., 2013) and falls under the biology of organizational behavior (Colarelli and Arvey, 2015). Genetic studies place more emphasis on genes and their roles in determining leadership behavior than context. For example, Chaturvedi et al. (2012) studied the heritability of emergent leadership and found that emergent leadership has genetic components, explaining 44 percent of the vari- ance for women and 37 percent for men. We must acknowledge that leadership behavior includes several other behaviors, such as decision making, cooperation, trust, change, and emotional regulation whose neural basis is also discussed in this book. Before discussing its neural foundations, it is important to present a brief overview of the nature of leadership in organizations. In doing so, the section summarizes the most popular leadership theories discussed in current organizational behavior textbooks.

1 THE NATURE OF LEADERSHIP

Research in social psychology and organizational behavior defines a leader as a person in a group who exercises the most influence over others. This definition implies that leadership is in fact a process of influence. If leadership is a process of influence, then there is no leader without followers. To be a leader, one should have people who voluntarily agree to be influenced. It is also important to note that leadership implies reciprocal influence because leaders and followers mutually influence each other. Hence, leadership is a two-way street. Leadership also takes place in a social context. Thus, an interactional framework is needed to better understand the leadership process.

1.1 Leadership as an Interactional Framework (Stogdill, 1948; Bass, 1985)

An interactional framework advocates that it is useful to study leadership as a process rather than to focus only on the leader as an individual. According to the interactional framework, to understand leadership we must assess the interplay between (1) leaders' characteristics, (2) followers' characteristics, and (3) situational factors.

When considering the leader, the focus is on his or her characteristics. These personal characteristics include personality traits, perceptions, and actual behaviors. The leader's personality may influence his or her reactions toward specific situations and followers. Examples of personality traits include emotional stability, assertiveness, risk proneness (or risk aversion), and open-mindedness. In addition to personality traits, the leader's cognitive abilities can also shape his or her behavior. Factors such as intelligence, creativity, and self-monitoring, to name but a few, can influence the leader's behavior. Cognitive abilities are also related to level of education and knowledge. Nevertheless, these characteristics of the leader alone are not enough to explain the leadership process. We need to consider followers' characteristics as well as the characteristics of the situation.

In discussing followers' characteristics, we find similar factors. Followers' personality and cognitive abilities can affect their behavior and relations with the leader. For example, followers who are open to experience are more likely to embrace change than those who are not. Intelligent, educated, and knowledgeable followers may require leadership styles that are different from those followers who do not possess such cognitive resources. When followers are highly educated and knowledgeable, there is often no need for an autocratic (or directive) leadership style. The number of followers a leader oversees can also impact the leader's behavior. Finally, we consider the followers' level of maturity – mature followers and immature followers would require different leadership styles. The key idea here is that followers' characteristics shape the different styles of leadership.

The context or situation means both a specific task and the social context in which leadership takes place. A situation may be simple (all parameters are known) or complex (some key parameters are unknown and/or numerous). Such characteristics affect both leaders' and followers' behavior. For instance, in a situation of downsizing, a leader may adopt a behavior that is quite different from that in a situation of performance appraisal. Likewise, in a situation of crisis, followers look to leaders for guidance and reassurance.

1.2 Contingency Theories of Leadership

Contingency theories of leadership contend that there is no best way to lead people. Indeed, leadership effectiveness depends on the context. Several contingency theories are briefly reviewed below.

1.2.1 Fiedler's contingency model

According to Fiedler (1964, 1967), the performance of a leader depends on two interrelated factors: (1) the degree to which the situation gives the leader control and influence, that is, the likelihood that the leader can successfully accomplish the job; and (2) the leader's basic motivation, that is, whether the leader's self-esteem depends primarily on accomplishing the task or on having close supportive relations with followers. Fiedler identified two classifications of leadership styles: relationship motivated and task motivated based on the Least Preferred Coworker (LPC) scale. Fiedler argues that leadership style is permanent and difficult to modify, therefore what needs to be modified is the situation. To be effective, leaders should work in situations that match their style. A key concept in Fiedler's contingency model is the leader-match concept. Effectiveness depends on matching leaders to situations. Task-motivated leaders are effective in situations of both high and low control, whereas relationship-motivated leaders are effective in situations of moderate control.

1.2.2 Path–goal theory

The path–goal theory of leadership (House, 1971) specifies what a leader must do to achieve high productivity and morale in each situation. It is an extension of previous theories of leadership that focused on consideration and initiative structure (Fleishman, 1953a). The leader must clarify the path, set goals, and provide information related to how to accomplish those goals. In any organization there are factors beyond the control of group members that influence task and satisfaction. These factors include other group members' tasks and the authority system within the organization. Effectively using path–goal theory requires assessing the relevant environmental variables and selecting one of four leadership styles: directive, supportive, participative, and achievement oriented.

1.2.3 Situational leadership model

Hersey and Blanchard (1969, 1977, 1982) developed the situational leadership model. This model considers two leadership dimensions: task behavior and relationship behavior. Task behavior refers to the extent to which the leader spells out the responsibilities of an individual or a group. Relationship behavior, however, refers to friendly relationships with group members. The relative effectiveness of these two dimensions often depends on the situation. The authors developed several combinations of these two dimensions. They believe that some combinations are more effective than others. They also considered the desirable characteristics of followers. Among these characteristics are job maturity and psychological maturity. Job maturity refers to the amount of task-relevant knowledge, experience, and skills. Psychological maturity refers to followers' self-confidence, commitment, motivation, and self-respect relative to the task.

To effectively use the situational leadership model, leaders should first assess followers' maturity level relative to the task. According to this model, leaders may implement a series of developmental interventions to help boost followers' maturity levels. This process would begin by assessing followers' current level of maturity and then determining the leader behavior that best suits that follower regarding the particular task. Leaders should assess followers' readiness. Followers' readiness includes ability and willingness to perform a given task. Based on this, the authors developed four leadership styles: the telling style, the selling style, the participating style, and the delegating style. The appropriateness of each style depends on followers' readiness.

1.2.4 Leader–member exchange model (LMX model)

According to this model, the type of exchange between the leader and followers determines group effectiveness (Graen, 1976). The model contends that leaders have special relations with their subordinates and these relationships shape the outcomes of the exchange. There are two group members: the "in-group" members and the "out-group" members.

There is more trust between the leader and in-group members. Those members receive better rewards, have frequent interactions with the leader, and develop high levels of performance and satisfaction. In contrast, out-group members have fewer interactions with the leader and their interactions are more formal. There is also less trust between the leader and the out-group members. These members have lower levels of satisfaction and performance and receive fewer rewards from the leader.

1.3   Contemporary Theories of Leadership

There are several theories of leadership that could be considered as modern or contemporary. However, in the following sections, I only focus on three of them – charismatic leadership, transformational leadership, and transactional leadership – as the nascent research on the neural basis of leadership has explored their neural foundations.

1.3.1   Charismatic leadership (Conger and Kanungo, 1987, 1988; Shamir, House, and Arthur, 1993)

Charismatic leadership is another way of saying that traits matter for effective leadership. Charisma refers to extraordinary abilities that followers attribute to the leaders. Charisma is not a physical trait that a leader possesses per se; it is an attribution that followers make about the leader. As beauty lies in the eye of the beholder, charisma lies in the eye of the follower. This charisma helps create an emotional bond between the leader and his or her followers. And because followers "like" the leader, they are likely to follow him or her and obey his or her orders. Leaders considered as charismatic are visionary, inspire trust, and possess excellent communication skills. Charismatic leaders also have strong convictions about the vision and show emotional expressiveness and warmth. They use unconventional strategies and are self-confident.

1.3.2   Transformational leadership

Whereas charismatic leadership deals with the leader's personal attributes, transformational leadership deals with what the leader does, what types of changes the leader brings to the group, the organization, or the community. Transformational leaders motivate others to work for the collective good by linking individual values with a values-based vision of the future and serve as role models (Bass, 1985; Bass and Avolio, 1995; Ramchandran et al., 2016). Transformational leadership includes four key characteristics: idealized influence (charisma), inspirational leadership, intellectual stimulation, and individualized consideration. By being charismatic, the transformational leader has a vision and a sense of a mission.

This helps him or her gain respect and loyalty from followers. By practicing inspirational leadership, the leader communicates his or her vision to followers. Transformational leaders use emotional appeals to inspire followers. By providing intellectual stimulation, the leader encourages followers to analyze problems in new ways, creates an atmosphere conducive to creative thinking, and emphasizes the use of methodical problem solving and reasoning. Finally, by providing individualized consideration, the leader pays personal attention to employees, uses a one-on-one communication style, and emphasizes their personal development.

1.3.3 Transactional leadership (Avolio, Bass, and Jung, 1999)

Transactional leadership is characterized by a contractual relationship between leader and followers. This contractual relationship may be explicit or implicit. An explicit contractual relationship is a relationship based on formal authority as spelled out in a formal contract such as that between an employee and his or her supervisor. An implicit contractual relationship is a de facto relationship between a leader and a follower that is not based on a written contract. The leader behavior is associated with constructive and corrective transactions. This transactional relationship implies that subordinates will work in exchange for rewards. A transactional leader can use a constructive style or a corrective style. When the leader uses a constructive style, rewards are contingent on performance – followers are rewarded when they accomplish their objectives. In a corrective style, the leader manages by exception, passively or actively. In managing by exception actively, the leader intervenes to criticize employees and focuses on mistakes. In managing by exception passively, the leader does not intervene unless he or she is required to do so. When the leader intervenes, he or she will point out mistakes and wrongdoings.

All the leadership models and theories discussed above could be assessed through the lens of neuroscience. A neuroscience perspective on leadership can include two dimensions. First, one can study the neural foundations of leadership by expanding the neural foundations of behaviors that are related to leadership, such as decision making, emotions, empathy, ethics, motivation, and many others. Indeed, leaders exercise behaviors whose neural foundations have been established. They make decisions, motivate employees, and express emotions. Second, organizational scholars can directly study the neural basis of leaders displaying leadership characteristics such as charismatic leadership, transformational leadership, transactional leadership or leaders showing considerate behavior toward their followers. The following section reviews the extant literature on the neural basis of leadership.

2 NEUROSCIENCE AND LEADERSHIP

2.1 Leadership and the Brain

For the proponents of a neuroscience perspective, understanding the brain can provide useful insights into improving knowledge about leadership and leadership development. Waldman et al. (2011a) discussed the role of neuroscience in understanding inspirational leadership. They used the construct of coherence in neuroscience, which refers to the extent to which two brain regions coordinate their activities: "coherence is a way of measuring the interconnectedness of areas of the brain" (Waldman et al., 2011a, p. 62). The authors found that right frontal coherence was associated with participants who were coded as high on socialized visionary communication ($r = 0.36$, $p < 0.05$) and socialized vision was correlated with follower perceptions of inspirational/charismatic leadership ($r = 0.39$, $p < 0.01$). Coherence is reported as a percentage, where a high percentage indicates high coherence and a low percentage indicates low coherence.

Hannah et al. (2013) studied a sample of 103 military leaders using quantitative electroencephalography (qEEG) and found that complex leaders showed more adaptability in decision making as indicated by the complexity of the tasks. A lower level of EEG coherence in the alpha frequency range was associated with greater adaptive decision making. This indicates that brain regions such as the ventromedial prefrontal cortex (vmPFC), the dorsolateral prefrontal cortex (dlPFC), and the anterior cingulate cortex (ACC) were activated. The vmPFC is involved in self-regulation, the dlPFC in attention processes, choice, and novelty processing, and the AAC in behavior monitoring and guiding. These regions intervene in the complexity of exercising leadership.

In another study of a sample of 200 civilian and military leaders using EEG measurements coupled with scores from the Multifactor Leadership Questionnaire (MLQ), Balthazard et al. (2012) found a neural basis for differentiating transformational and non-transformational leaders. Areas of the brain that were activated included the frontal lobes, the central and parietal lobes, the temporal lobes, and the parietal lobes. The authors concluded that transformational leadership had a "neural signature."

These findings were consistent with previous results in cognitive neuroscience. Neurological variables may help provide a better understanding of why leaders do what they do (Waldman et al., 2011b, p. 1092).

In a study conducted on a sample of 105 mid-level and senior managers, Ramchandran et al. (2016) found that the two dimensions of executive function – control (inhibition of pre-potent response, flexible thinking) and decision making – interacted to predict transformational leadership, controlling for the extant antecedents of extraversion and general mental ability. Specifically, transformational leadership was associated with (1) high inhibition of pre-potent response in the presence of low-risk decision making; and (2) either mental flexibility or low-risk decision making, interchangeably.

Rochford et al. (2017) explored the implications of opposing domains theory for developing ethical leaders. Opposing domains theory highlights a neurological tension between analytic reasoning and socioemotional reasoning. The authors contend that when people engage in analytic reasoning (the task-positive network), they suppress their ability to engage in socioemotional reasoning (the default mode network) and vice versa. Rochford et al. (2017) propose that a key issue for ethical leadership is achieving a healthy balance between analytic reasoning and socioemotional reasoning. If there is imbalance, leaders run the risk of suppressing their ability to pay attention to the human side of moral dilemmas and, in doing so, dehumanize colleagues, particularly subordinates, and clients.

Waldman, Wang et al. (2017) found that the interaction of leader relativism and idealism partially mediates the effects of the brain's default mode network in the prediction of ethical leadership. Relativism assumes that ethical principles are context dependent. Because people act and react based on the contingencies they face, relativism contends that ethical principles could differ from context to context. This view of ethics is contrary to universalism, which assumes that there are universal ethical principles that can be applied to all contexts. Idealism is an approach to ethics where people are willing to act in such a way as to not harm others. A person who is idealistic strives to act ethically to avoid causing prejudice to others.

2.2 Theory of Mind, Mindfulness, and Leadership

Two emerging concepts, mindfulness and theory of mind (or mentalizing), are important for developing today's leaders. Mindfulness refers to the extent to which one is aware of one's inner self and understands one's feelings and cognitions. Being aware of one's self could prove useful in understanding one's strengths as well as weaknesses. Theory of mind or mentalizing refers to the capacity to acknowledge that others have feelings and cognitions that drive them. According to Premack and Woodruff (1978), an individual has a theory of mind if he or she imputes mental states to themselves and others (p. 515). Theory of mind is the "ability to explain and predict the behavior of ourselves and others by attributing to them independent mental states, such as beliefs, desires, emotions or intentions" (Gallagher and Frith, 2003, p. 77). In the context of leadership, leaders should understand themselves and others.

The three areas consistently activated in association with theory of mind include the anterior paracingulate cortex, the superior temporal sulci, and the temporal poles bilaterally. Mindfulness practice is associated with neuroplastic changes in the anterior cingulate cortex, insula, temporoparietal junction, frontolimbic network, and default mode network structures (Hölzel et al., 2011). Ives-Deliperi, Solms, and Meintjes (2011) employed functional magnetic resonance imaging (fMRI) to identify the brain regions involved in the state of mindfulness and to shed light on its mechanisms of action. Significant signal decreases were observed during mindfulness meditation in midline cortical structures associated with interoception (the signaling and perception of internal bodily sensations), including bilateral anterior insula, left ventral anterior cingulate cortex, right medial prefrontal cortex, and bilateral precuneus. Significant signal increase was noted in the right posterior cingulate cortex.

Understanding others could help leaders become more effective. Both cognitive elements, mindfulness and mentalizing, have neural foundations. For example, efforts to understand one's self and others can create self-directed neuroplasticity, which depends more on conscious efforts. A person may deliberately focus on specific behaviors and actions. Consequently, the brain may "rewire" itself toward these activities. Neuroplasticity refers to the brain's ability to rewire itself. The late Canadian psychologist Donald Hebb (1949), considered the father of neuroplasticity, argues that neurons that fire together wire together. In other words, parts of the brain that are continually activated together will physically associate with one another in the future.

The two concepts of mindfulness and theory of mind are important in understanding the neural basis of leadership. Theory of mind helps anticipate the actions of others (Gallagher and Frith, 2003) – leaders are able to anticipate followers' behavior. This knowledge can help them design strategies to successfully interact with followers and motivate them. Likewise, mindfulness could be useful to leaders by allowing them to regulate and reappraise their emotional states.

2.3 Neural Basis of Resonant and Dissonant Leadership

According to Boyatzis and McKee (2005), resonant leaders are in tune with those around them. "Leaders who can create resonance are people who either intuitively understand or have worked hard to develop emotional intelligence, namely, the competencies of self-awareness, self-management, social awareness, and relationship management" (Boyatzis and McKee, 2005, p. 4). On the other hand, dissonant leaders are not in tune with their environment and are at the mercy of volatile emotions and frustrate and stress their followers.

Boyatzis et al. (2012) conducted an fMRI study to localize brain regions involved in resonant and dissonant leadership. They described "resonance as physiological attunement and interpersonal synchrony between a leader and another individual and dissonance, as a lack thereof" (Boyatzis et al., 2012, p. 261). The authors had participants recall incidents with both types of leaders. They found that recalling experiences with resonant leaders activated the bilateral insula, the right inferior parietal lobe, and the left superior temporal gyrus. These regions are often associated with the mirror neuron system, default mode network, social network, and positive affect. However, recalling experiences with dissonant leaders negatively activated the right anterior cingulate cortex, the right inferior frontal gyrus, and bilateral frontal gyrus/insula. These regions are associated with the mirror neuron system and relate to avoidance, narrowed attention, decreased compassion, and negative emotions. Boyatzis et al. (2012) contend that "relationships with resonant leaders are characterized by mutual positive emotions, a subjective sense of being in synchrony with one another, and physiological effects of parasympathetic nervous system activation. In contrast, relationships with dissonant leaders produce negative emotions, interpersonal discord, and sympathetic nervous system activation (e.g., flight-or-fight response" (p. 261). In fact, "[r]esonance and dissonance emerge in leader relationships through the exchange of emotions" (ibid.).

Jack, Dawson, and Norr (2013) studied the neural basis of the two types of coaching: positive emotional attractor (PEA) and negative emotional attractor (NEA). PEA emphasizes compassion for the individual's hope and dreams, whereas NEA focuses on externally defined criteria. Using fMRI, the authors found that PEA leads to the activation of lateral occipital cortex, superior temporal cortex, medial parietal, subgenual cingulate, nucleus accumbens, and the left prefrontal cortex. These activations are related to visioning, parasympathetic nervous system activity, and positive affect. Regions showing more activity in the NEA condition included medial prefrontal regions and the right lateral prefrontal cortex. These activations were related to sympathetic nervous system activity, self-trait attribution, and negative affect.

Peterson et al. (2008) studied the neural basis of psychological capital (PsyCap – optimism, hope, confidence, and resilience). They found that "hopeful, optimistic, confident, and resilient individuals were persistent and disciplined goal setters, and remained positive in the face of adversity" (p. 347). These qualities are essential to effective leadership. The authors built brain maps of leaders and compared them using EEG technology. They found that high PsyCap participants tended to show greater activity in the left, prefrontal cortex than the low PsyCap participants (ibid.). In contrast, low PsyCap leaders showed more activity in the right frontal cortex and the right amygdala (ibid.). The authors conclude that leaders who engage in relational practices and foster a sense of interdependence among people will have a positive effect on their subordinates' physiological resourcefulness. This is important because positive social interactions have beneficial effects, including better health (Heaphy and Dutton, 2008).

2.4 Neuroscience and Leadership Development

Can knowledge from neuroscience help develop better leaders? This question is relevant and has been explored by authors such as Waldman et al. (2011a, 2011b) who considered that neuroscience can lead to leadership development. Indeed, Waldman et al. (2011a, 2011b), Balthazard et al. (2012), and Waldman and Balthazard (2015) have all emphasized the use of neuroscience to develop better leaders. For example, Waldman et al. (2011a, 2011b) and Balthazard et al. (2012) contend that neuroscience techniques such as neurofeedback can be used to develop transformational and inspirational leaders. Neurofeedback is a self-regulation technique providing individuals with feedback about specific levels of brain activity in conjunction with specific target behaviors (Waldman et al., 2011a, 2011b; Waldman and Balthazard, 2015; Lindebaum, Al-Ahmoudi, and Brown, 2017). By becoming aware of how their brains function, such leaders would be able to develop the skills required to become better leaders. Brain neuroplasticity has been considered as one of the ways in which neuroscience can contribute to leadership development. Waldman et al. (2011a) suggest that "if the neural pathways in the brain are malleable, there could be some interesting implications for leadership development" (p. 68).

Peterson et al. (2008) proposed a neurological approach to leadership development, including the following steps:

1. Access PsyCap in leaders. According to the authors, EEG data can help to identify hope, optimism, confidence, and resiliency. Hence, organizations can use neuroscience to identify these traits in leaders.
2. Identify and address neurological barriers to PsyCap in leaders. EEG data can help to identify conditions that may prevent leaders from displaying the four components of PsyCap.
3. Develop PsyCap leaders. To develop PsyCap leaders using a neurological approach the authors suggest that organizations do the following: (a) combine neurofeedback with coaching; (b) keep the motivation climate high; (c) include meditation as part of employee wellness; (d) keep the training short and sweet; (e) use cross-functional PsyCap; (f) encourage innovation; and (g) reward positivity.

Nevertheless, although this approach provides some useful guidelines for developing leaders, one could argue that we do not necessarily need a neuroscientific approach to make these recommendations. In fact, training programs that focus on these areas do not always rely on neuroscience.

3 NEUROSCIENCE AND FOLLOWERSHIP

As stated at the beginning of the chapter, leadership is a two-way street and requires the will of followers to voluntarily follow the leader. Hence, the study of the neural basis of leadership should also include a dimension on followership. Vugt and Ronay (2014) propose that leadership and followership evolved in humans and in other species to solve recurrent coordination problems. They define leadership in terms of the coordination of the actions of two or more individuals to accomplish joint goals.

3.1 Neural Evidence of Leader–Follower Interactions

Neuroscientific evidence could help to explore followers' reactions to inspirational and non-inspirational messages from leaders. In a study designed to assess how followers' brains respond to such messages, Molenberghs et al. (2015) conducted an fMRI study on the followers of inspirational leaders and found that inspirational statements from in-group leaders were associated with increased activation in the bilateral rostral inferior parietal lobe, pars opercularis, and posterior midcingulate cortex. These brain regions are often implicated in semantic information processing. In contrast, for out-group leaders, greater activation in these areas was associated with non-inspirational statements. Non-inspirational statements by in-group leaders resulted in increased activation in the medial prefrontal cortex (mPFC). The mPFC is often associated with reasoning about a person's mental state. In this study, participants engaged in greater semantic processing of (1) inspirational collective-oriented messages from in-group leaders; and (2) non-inspirational personal-oriented messages from out-group leaders. This indicates that participants processed information that was aligned with their existing beliefs. Hence, followers may tend to react more positively to leaders they associate with more.

One of the tasks of leaders is to motivate and reward employees, hence, it is important for leaders to understand the role of neuroscience in human motivation. Such understanding could help to improve leaders' effectiveness in managing their followers. Zink et al. (2008) investigated the neural basis of social hierarchy in humans and found that in both stable and unstable social hierarchies, viewing a superior individual differentially engaged perceptual-attentional, saliency, and cognitive systems, notably the dorsolateral prefrontal cortex. However, in the unstable social hierarchy setting, brain regions related to emotional processing such as the amygdala and the mPFC were activated. These findings have implications for the study of the neural basis of leadership in organizations.

Jiang et al. (2015) investigated whether interpersonal neural synchronization (INS) played an important role in leader emergence, and whether INS and leader emergence were associated with the frequency or the quality of communications. Eleven three-member groups were asked to perform a leaderless group discussion task, and their brain activities were recorded via functional near-infrared spectroscopy-based hyperscanning. Video recordings of the discussions were coded for leadership and communication. The results showed that the INS for the leader–follower pairs was higher than for the follower–follower pairs in the left temporo-parietal junction, an area important for social mentalizing.

Schjoedt et al. (2010) used fMRI to investigate how assumptions about speakers' abilities changed the evoked blood oxygenation level-dependent (BOLD) signal response in secular and Christian participants who received intercessory prayer. They found that recipients' assumptions about senders' charismatic abilities have important effects on their executive network. Most notably, the Christian participants deactivated the frontal network consisting of the medial and the dorsolateral prefrontal cortex bilaterally in response to speakers who they believed had healing abilities. An independent analysis across subjects revealed that this deactivation predicted the Christian participants' subsequent ratings of the speakers' charisma and experience of God's presence during prayer. These findings could have implications for leadership in organizations. Leaders perceived as charismatic can ignite a sense of commitment to organizational goals and motivate employees to perform at a higher level.

3.2 Followership and Emotional Contagion

Hatfield, Cacioppo, and Rapson (1994) pointed out that one person's affective state can influence that of others. This is particularly important for leadership where the affective state of a leader can influence followers. Dasborough (2006) used affective events theory (AET) as a conceptual framework to argue that leaders are sources of employee positive and negative emotions at work. AET assumes that emotions in organizations are triggered by specific events. In the case of leadership, effective leaders shape the affective events that influence the attitudes and behaviors of employees in the workplace (Weiss and Cropanzano, 1996). Certain leader behaviors displayed during interactions with their employees are the sources of these affective events.

The second theoretical underpinning of this theory is the asymmetry effect of emotion. Consistent with this effect, employees are more likely to recall negative incidents than positive ones. In a qualitative study, Dasborough (2006) found evidence that these processes exist in the workplace. Leader behaviors were sources of positive or negative emotional responses in employees, but employees recalled more negative incidents than positive incidents, and they recalled them more intensely and in more detail than positive incidents. This asymmetry effect could play an important role in the interactions between leaders and followers. Leaders must be mindful that negative experiences with followers are more likely to leave lasting impressions than positive ones.

Lee (2015) contends that there are brain-related processes that lead to emotional contagion. Lee also discussed the construct of emotional convergence. However, leaders may not always be willing to share the emotions of their followers. This could lead to "compassion fatigue." Emotional contagion is tacit and subconscious. However, it can be explicit, for example, in interpersonal relationships such as between leaders and followers. The mirror neuron system "creates an instant sense of shared experience" (Goleman and Boyatzis, 2008, p. 3). People have known that the way a message is delivered is often more important than the message itself. Hence, leaders should pay particular attention when they provide feedback to their followers because strong negative emotions tend to be more vivid than weak negative ones.

3.3 Challenges of Neuroscientific Studies of Leadership

Despite the potential of neuroscience to advance our understanding of leadership, several authors have noticed that the use of neuroscientific methods should not be considered as a panacea. Echoing this, Lee, Senior, and Butler (2012a) argue that a neuroscience approach should not overlook the importance of other methods that have helped shape current knowledge on leadership. Although Waldman et al. (2011b) contend that neuroscience could provide new insights into theory building and leader development, they caution researchers on the complexity of leadership because its understanding involves the leader, followers, and the context.

Lindebaum and Zundel (2013) contend that reductionism could limit the appeal of neuroscientific approaches to the study of leadership. This argument suggests that leadership behavior, and for that matter human behavior, is complex and cannot be reduced to neural processes. Hence, Lindebaum and Zundel (2013) maintain that "an unbridled pursuit of organizational neuroscience is more likely to impoverish rather than enrich our substantive understanding of leadership" (p. 859). Indeed, "human behavior is essentially relational and the brain level is not always the ultimate cause of human behavior, but merely one part of more complexly unfolding processes" (p. 872). This is particularly true in the domain of leadership.

Lindebaum (2013a) argues that some of the claims made by organizational neuroscientists could pose ethical and moral dilemmas. For example, is it possible to use neuroscientific interventions to alter people's behavior in organizations? Is it ethical to assume based on neuroscience data that some leaders are "deficient" and therefore require some form of intervention? Doing so would imply that these leaders are "abnormal" and therefore a neuroscientific intervention is needed (neuroscientific therapy). Lindebaum et al. (2017) point to neurofeedback as a case in point. The authors contend that the use of neurofeedback is not likely to improve leadership development. They also suggest that it could raise ethical issues for at least two reasons. It may be assumed that the results are valid and reliable when in fact they are no different than those obtained from conventional measures of leadership. Second, it may lead people to consider that some leaders are "neurologically deficient" when in fact there is no reason to suspect they are.

We must acknowledge that not all authors are skeptical about the use of neuroscience in understanding leadership. In fact, some authors have questioned the criticism of the use of neuroscience evidence in organizational behavior and leadership. For example, Ashkanasy (2013) uses his personal experience as a scholar who previously worked on the concept of emotional intelligence that was then considered as a management fad, to caution critics of neuro-leadership "not to throw the baby out with the bathwater." Although Lafferty and Alford (2010) suggest that a neuroscientific approach offers a new avenue for overall learning, growth, and change, they mention that it is too early to decide whether such an approach offers the "kind of validation that will help sustain the study of leadership and organizational management" (p. 37). We must acknowledge that "theory and research attempting to connect neuroscience and leadership are still at a nascent stage of development" (Balthazard et al., 2012, p. 246). Hence, there is room for refinement of theories and methods.

7. The neural basis of fairness

Fairness is one of the key concerns when people become members of organized groups such as organizations, families, villages, communities, and nations, and has been studied by several disciplines such as economics, psychology, philosophy, and sociology. As Sanfey et al. (2003, p. 1757) put it, "a basic sense of fairness and unfairness is essential to many aspects of societal and personal decision making and underlies notions as diverse as ethics, social policy, legal practice, and personal morality." Pressman (2006) goes even further to argue that we do not learn to be fair because fairness is part of our genetic makeup.

People have a fundamental sense of fairness that represents a motivational source when they make decisions (Rabin, 1993; Karni and Safra, 2002a, 2002b; Karni, Salmon, and Sopher, 2008). In the workplace, fairness is an important "yardstick that employees use to assess outcome distribution, formal procedures, or interpersonal treatment in organizations" (Beugré, 2009, p. 129). Karni and Safra (2002b) measured what they called the "intensity of the sense of justice" and found that this intensity is stronger in some people than in others. Some people may experience an intense sense of gratification when they act fairly (or experience fairness) or suffer more when they act unfairly (or witness unfairness). Karni and Safra (2002b) concluded that the intensity of the sense of fairness influences people's behavioral choices. Schmitt, Neumann, and Montada (1995) introduced the construct of justice sensitivity to explain the degree to which people react to fair or unfair treatment in general. Justice sensitivity includes three dimensions: (1) sensitivity toward experiencing injustice toward oneself; (2) sensitivity to observing that others are treated unfairly; and (3) sensitivity to profiting from unfair events (Fetchenhauer and Huang, 2004).

The neuroscientific study of fairness can bring together economists, social psychologists, organizational scholars, sociologists, and anthropologists. The concept of fairness has been studied by economists (Rabin, 1993; Fehr and Schmidt, 1999; Fehr and Gächter, 2000) and neuroeconomists (Camerer et al., 2004, 2005; Camerer, 2007) but these scholars have completely ignored the work of organizational justice scholars (Beugré, 2009). Likewise, organizational justice scholars have studied fairness in organizations for more than three decades but they too have also ignored the work of economists and neuroeconomists. Hence, a neuroscientific perspective could help to build a bridge between these groups of scholars. This would require that neuroeconomists pay attention to the "real-life" applications of laboratory studies on the neural basis of fairness and that organizational justice scholars familiarize themselves with technical developments in the neural studies of fairness. The pairing of experimental gaming and neuroscience techniques has unique value for organizational justice researchers (Volk and Köhler, 2012) – such an approach could lead to a mutual enrichment of both disciplines.

1 NEURAL BASIS OF REACTIONS TO FAIRNESS

Research in neuroeconomics has found that the anterior insula is activated when facing situations leading to decisions of fairness or unfairness and to emotional responses such as pain and disgust (Sanfey et al., 2003). Knoch, Pascual-Leone, and Myer (2006) showed that disruption of the right but not the left dorsolateral prefrontal cortex (dlPFC) by low-frequency repetitive transcranial magnetic stimulation (rTMS) substantially reduces participants' willingness to reject their partners' intentionally unfair offers in the Ultimatum Game although they still judge such offers as unfair. In light of these findings, Knoch et al. (2006) conclude that the right dlPFC plays a key role in the implementation of fairness-related behaviors. Fairness also activates the ventromedial prefrontal cortex (vmPFC) (McCabe et al., 2001), whereas unfairness activates the dorsomedial prefrontal cortex (dmPFC) (Decety et al., 2004). Tabibnia, Satpute, and Lieberman (2008) found that the orbitofrontal cortex (OFC) was associated with fairness preferences.

Research on the neural basis of fairness has used several methods, including functional magnetic resonance imaging (fMRI), positron emission tomography (PET), electroencephalography (EEG), and transcranial magnetic stimulation (TMS). The conclusion is that certain brain regions are activated when participants experience fairness or unfairness. Studying the neural basis of fairness could lead to the development of a brain map of fairness: a brain map is "a visual image of the brain which facilitates the neuroscientist's ability to compare one brain to another" (Peterson et al., 2008, p. 346). Instead of a picture of the human brain, I present here a table that lists some of the brain regions implicated in reactions to fairness and unfairness (Table 7.1).

Examples of brain regions activated by perceptions of fairness include the anterior insula (AI), the nucleus accumbens, the anterior cingulate cortex (ACC), the orbitofrontal cortex (OFC), the lateral prefrontal cortex (lPFC), and the superior temporal sulcus (STS). Brain regions activated by perceptions of unfairness include the anterior cingulate cortex (ACC), the insula, the dorsolateral prefrontal cortex (dlPFC), the dorsomedial prefrontal cortex (dmPFC), the medial prefrontal cortex (mPFC), the superior temporal sulcus (STS), the ventral striatum (VS) and the ventrolateral prefrontal cortex (vlPFC).

Table 7.1 Brain regions associated with fairness and unfairness

Regions of Interest (ROI) for Fairness	Regions of Interest (ROI) for Unfairness
Anterior insula (AI)	Anterior cingulate cortex (ACC)
Nucleus accumbens (NAcc)	Insula (I)
Orbitofrontal cortex (OFC)	Dorsolateral prefrontal cortex (dlPFC)
Anterior cingulate cortex (ACC)	Dorsomedial prefrontal cortex (dmPFC)
Lateral prefrontal cortex (lPFC)	Medial prefrontal cortex (mPFC)
Superior temporal sulcus (STS)	Superior temporal sulcus (STS)
	Ventral striatum (VS)
	Ventrolateral prefrontal cortex (vlPFC)

Unfairness tends to activate brain regions that are also implicated in pain and disgust, whereas fairness tends to activate brain regions that are implicated in reward. For example, fMRI studies showed that unfair offers activated brain regions related to emotions (anterior insula) as well as cognition (dorsolateral prefrontal cortex). There is also an increased activity in the anterior insula when participants rejected unfair offers, thereby suggesting an important role of emotions in decision making. Unfair offers that are rejected have greater anterior insula activation than dorsolateral prefrontal cortex activation, whereas accepted offers exhibited greater dorsolateral prefrontal cortex (dlPFC) activation than anterior insula activation (Sanfey et al., 2003). "The activation of the dlPFC during unfair offers may be due to the fact that an unfair offer is more difficult to accept as indicated by the higher rejection rates of these offers, and hence higher cognitive demands may be placed on the participants in order to overcome the strong emotional tendency to reject the offer" (Sanfey et al., 2003, p. 1757).

Unfair offers also activate the anterior cingulate cortex (ACC), which has been associated with detection of cognitive conflict, thereby indicating a cognitive conflict between cognitive and emotional motivations in the Ultimatum Game (Sanfey et al., 2003). This activation of the ACC could also indicate a conflict between accepting and rejecting an unfair offer. Scholars tend to agree that unfair offers elicit negative emotions, such as anger and frustration, which in turn lead to rejection decisions (Pillutla and Murnighan, 1996). Pillutla and Murnighan (1996) even suggest that feelings of anger are a better predictor of rejection of unfair offers than the unfairness of the offers themselves.

Lamichhane et al. (2014) use fMRI experiments to study refusals and protests using both favoring and disfavoring inequity in three economic exchange games with undercompensating, nearly equal, and over-compensating offers. Refusals of undercompensating offers recruited a heightened activity in the right dlPFC. Accepting overcompensating offers recruited significantly higher node activity in, and network activity among, the caudate, the cingulate cortex, and the thalamus. Protesting undercompensating fixed offers activated the network consisting of the right dlPFC and the left ventrolateral prefrontal cortex (vlPFC) and midbrain in the substantia nigra.

Wout et al. (2006) observed 30 undergraduate students play the Ultimatum Game while their skin conductance responses were measured as an automatic index of affective state. Skin conductance activity was higher for unfair offers and was associated with the rejection of unfair offers. In other words, participants experience more emotional arousal when confronted with an unfair offer compared to a fair offer (Wout et al., 2006). However, this pattern was observed for unfair offers made by humans and not by computers.

1.1 Social Distance and Responders' Reactions

Does social distance affect responders' reactions to fairness and unfairness? This is an empirical question that several authors have addressed. For example, Wu, Zhou et al. (2011) found that event-related brain potentials recorded from participants showed that highly unequal offers elicited more negative-going medial frontal negativity than moderately unequal offers. Similar results were also found by Wu, Leliveld, and Zhou (2011) but the medial frontal negativity was more negative for offers made by friends than by strangers. Nevertheless, Campanha et al. (2011) did not support Wu and colleagues' results. They found that when the proposer was a friend rather than an unknown person, unfair offers were less frequently rejected and the medial frontal negativity, typically associated with unfair offers, was reversed to positive polarity. In other words, unfair offers from friends elicited a significant positivity, whereas unfair offers from unknown proposers elicited a significant negativity.

These studies indicating that reactions to unfairness relate to the nature of the offender (a friend or a stranger) have produced mixed results. On one hand, we may conjecture that people may expect more fairness from friends, and unfair treatment from friends would arouse stronger emotional feelings than unfairness from strangers. Unfair treatment from friends may convey a stronger sense of betrayal in that friends are supposed to look after friends. Conversely, people may be more willing to give the benefit of doubt to friends in the case of unfair treatment rather than to strangers.

1.2   Fairness as a Form of Reward

Neuroscientific research shows that fairness tends to activate the brain's reward circuitry. In this regard, fairness is perceived as a reward in itself. Knoch et al. (2006) showed that disruption of the right but not the left dlPFC by low-frequency rTMS substantially reduces participants' willingness to reject their partners' intentionally unfair offers in the Ultimatum Game. However, the participants to the experiment still judged such offers as unfair. Interpreting these findings, the authors conclude that the right dlPFC plays a key role in the implementation of fairness-related behaviors.

Most neuroscientific studies construe fairness as outcome related (Dulebohn et al., 2009) and used the Ultimatum Game to assess it. These studies also indicate that fairness activates the brain's reward circuitry, whereas unfairness activates brain regions implicated in painful reactions. Dulebohn et al. (2009) concluded that neuroscience research has approached fairness exclusively in terms of reactions to outcomes. One possible reason for these studies to construe fairness in terms of distributive justice could be the use of the Ultimatum Game as the main task that participants perform. In the Ultimatum Game, participants are often asked to decide whether to accept or reject offers made by proponents.

In this regard, the Ultimatum Game focuses on the fairness of outcomes distribution, which corresponds to distributive justice in the organizational justice literature. The conclusions that one could draw from these studies is that fairness has rewarding properties because it activates the brain's reward circuitry, such as the nucleus accumbens, the orbitofrontal cortex, the dorsal striatum, and the caudate nucleus. In the Ultimatum Game, fair offers have rewarding properties over and above the reward associated with receiving money (Tabibnia et al., 2008).

Because of its rewarding properties, fairness can be used as a motivational tool in organizations. This is particularly important for managers because acting fairly does not require the use of monetary incentives.

1.3   Reactions to Unfairness

Contrary to fairness, unfairness is treated as "painful" in the brain because it activates brain regions associated with reactions to pain and loss, such as the amygdala and the insula. Camerer et al. (2004) contend that experiencing unfairness may be seen as an unpleasant event that activates the insula, which is the wellspring of social emotions such as lust, disgust, pride, humiliation, guilt, and atonement. Thus, unfairness triggers activation of brain regions that process emotional reactions. The activation of the amygdala indicates that unfairness triggers emotional reactions in the brain. The same could be true for the anterior insula, which is also activated when facing situations leading to decisions of fairness or unfairness and to emotional responses (Sanfey et al., 2003; Morse, 2006). The insula processes information from the nervous system about bodily states, such as physical pain, hunger, the pain of social exclusion, disgusting odors, and choking (Camerer et al., 2004). Güroğlu et al. (2011) conducted an fMRI study on a sample of preteens and teens aged 10, 13, 15 and 20 respectively. They found that participants of all ages showed activation in the bilateral insula and dorsal anterior cingulate cortex (dACC) during rejection of unintentional but acceptance of intentional unfair offers.

Knowing that the experience of fairness is rewarding to the brain, whereas the experience of unfairness is painful, is an important finding for organizational justice scholars who strive to understand issues of fairness and provide guidelines to managers on how to display fair behaviors in the workplace. If fairness is rewarding, then organizational scholars have powerful scientific evidence with which to convince managers of the importance of acting fairly. Likewise, organizational justice scholars could also demonstrate to managers that unfairness is painful in the brain. Although we know that people resent being treated unfairly, scholars had assumed that this was more cultural than biological. In general, people prefer to be treated fairly and resent any form of unfairness experienced by themselves or others. Indeed, the benefits of feeling fairly treated in the workplace are well documented (Colquitt et al., 2001; Cropanzano and Wright, 2011).

Seymour, Singer, and Dolan (2007) found that the medial orbitofrontal cortex and the nucleus accumbens were activated when cues indicated an

imminent high-intensity shock to unfair players, and this activity correlated with individuals' subjective feelings of anger and retribution. The
authors concluded that the assumption that "punishment, including altruistic punishment, substantially promotes cooperation in human societies
seems firm" (Seymour et al., 2007, p. 307). Indeed, punishment is a
norm-maintaining mechanism and is essential to maintaining cooperation
and existing norms (Fehr and Fischbacher, 2004).

Dulebohn et al. (2009) conducted an fMRI study and concluded that
distributive justice and procedural justice activated different regions of
the brain. In a more recent study, Dulebohn et al. (2016) found evidence
that gender and distributive injustice interact to influence bargaining
behavior, with females rejecting Ultimatum Game offers more frequently
than males. They also showed that activation in the ventromedial
prefrontal cortex (vmPFC) and the ventral striatum brain regions during
procedural justice evaluation was associated with offer rejection in
females, but not in males. Although Dulebohn and his colleagues' studies
helped us understand the neural bases of two critical dimensions of
organizational justice, they did not cover interactional justice, which
deals with perceptions of fairness in the form of interpersonal treatment
with respect and dignity. It would be interesting to reveal brain regions
that might be activated by this justice dimension that involves social
interactions with others, such as superiors and peers. Since social
interactions can evoke strong emotions (Crockett et al., 2008), one could
speculate that interactional justice would be likely to activate brain
regions involved in emotional arousal.

1.4   Nature of Proponents and Reactions to Unfair Offers

Wout et al. (2006) found that skin conductance was higher when
participants received unfair offers from humans rather than computers.
These skin conductance responses could be construed as an expression of
emotional arousal. Hence, they could be interpreted as a signal that there
was a human target to retaliate against – that participants assigned
motives and intentions to the humans rather than the machines. Expanding this line of reasoning to potential neural reactions to interactional
justice, one could speculate that the experience of interactional injustice
could also lead to greater skin conductance activity in an EEG study and
activation of brain regions involved in emotions, such as the anterior
insula and the amygdala.

An interesting question that Wout et al.'s (2006) research raises is: why
do people tend to get angrier and reject unfair offers made by humans
rather than computers? One possible explanation is the preservation of

social status. The desire to preserve social status by rejecting an unfair offer is stronger when the proponent is another human (Nowak, Page, and Sigmund, 2000). It is also possible that reciprocity plays an important role when deciding to retaliate against a human offender rather than a machine. Knoch et al. (2006) used rTMS as a tool for studying participants' reactions to fairness and found that the dorsolateral prefrontal cortex (dlPFC) plays an important role in reciprocity. In this study, the disruption of the dlPFC did not affect reactions to unfair offers from computers. This could be explained by the fact that there was no opportunity to retaliate. In addition, one can hold grudges against another human but not against a machine.

These findings indicate that the sense of fairness is rooted in emotional processing (Hsu, Anen, and Quartz, 2008). Hsu et al. (2008) showed that disruption of the right but not the left dlPFC by low-frequency rTMS substantially reduces subjects' willingness to reject their partners' intentionally unfair offers. However, participants still judge the offers unfair. The right dlPFC plays a key role in the implementation of fairness-related behaviors. These results were observed when the proposer was a human rather than a computer. The dlPFC is crucial in overriding fairness impulses when self-interest and fairness motives are in conflict. Knoch et al. (2008) applied transcranial direct current stimulation (tDCS) to participants and showed that their propensity to punish unfair behavior was significantly reduced.

2 NEURAL BASIS OF THIRD PARTY REACTIONS TO UNFAIRNESS

Fairness is an important aspect of social interactions, so much so that people are willing to sacrifice personal gains to punish unfair behavior from others (see Fehr and Gächter, 2002; Camerer, 2003). Singer et al. (2006) found that men expressed a stronger desire for revenge than women and such desire for revenge activated the reward circuitry of the brain including the ventral striatum, the nucleus accumbens, and the orbitofrontal cortex. These findings support the view that humans derive satisfaction by simply seeing justice administered (Fehr and Gächter, 2002; De Quervain et al., 2004). They could also explain the desire for people to seek revenge when they feel unfairly treated. Crockett (2009) contends that rejecting unfair offers is antisocial at the individual level because it harms the proposer, depriving them of a monetary reward. However, it could be construed as a prosocial behavior at the group level because it enforces fairness norms (Knoch and Fehr, 2007; Crocket, 2009).

2.1 Altruistic Punishment

In a study using EEG, offers deviating from the equal division of assets elicit more negative-going medial frontal negativity (MFN) than equal offers in economic exchange games (Wu, Zhou et al., 2011). "Behavioral responses to social norm violation require at the very least, attention to social cues, sensitivity to change, and behavioral inhibition" (Watson and Platt, 2006, p. 186). In fact, human beings are sensitive to social norms. Any deviation from these norms creates a sense of unfairness and individuals are generally motivated to redress such wrongdoing by reacting emotionally or expressing moral outrage (Bies and Moag, 1986). Güroğlu et al. (2011) identified a norm violation network in the brain including the anterior insula and the dorsal cingulate cortex, and a social brain network including the dorsolateral prefrontal cortex and the temporoparietal junction. Generally, people have an inequity aversion (Fehr and Schmidt, 1999).

Norm violation is a fundamental concern in social interactions. Punishing norm violators has been construed as fair and a means not only to redress injustice but also to insure the smooth functioning of social institutions (Strobel et al., 2011). Strobel et al. (2011) studied reactions to violators of social norms and found that the reward regions of the brain, specifically the nucleus accumbens, were activated when norm violators were punished. This type of punishment is referred to as altruistic punishment. "Altruistic punishment is the costly punishment of norm violations, which does not involve any overt benefit for the punisher" (Strobel et al., 2011, p. 671). Wright et al. (2011) identified an inequality aversion network including the posterior insula, and Corradi-Dell'Acqua et al. (2012) found that the anterior insula is activated as a reaction to unfairness to self or a third party.

People often restrain their self-interest and behave altruistically to uphold principles of fairness (Lakshminarayanan and Santos, 2009) and care about the plight of others (Alexopoulos et al., 2012). Normative behaviors, such as volunteering, helping strangers, reciprocating favors, or punishing violators of social norms, are behaviors that people engage in to uphold social norms (Baumgartner et al., 2011). Such behaviors are essential for the smooth functioning of social institutions. Baumgartner et al. (2011) found that the activation of right dorsolateral prefrontal cortex (dlPFC) and the posterior ventromedial prefrontal cortex (pvmPFC), and the connectivity between them, facilitates subjects' willingness to incur the cost of normative decisions. Neural activation in the right dlPFC during the processing of unfair offers is decisive in the ability to make costly normative decisions.

The motivation to punish offenders could be partly due to feelings of satisfaction when social norm violations are punished and justice is restored (De Quervain et al., 2004). Behavioral reactions to unfairness are strongly modulated by the ascription of intentionality. Indeed, people react less negatively to disadvantageous inequity when they feel the inequity was not intentional (Blount, 1995). According to Alexopoulos et al. (2012), social preferences, such as inequality aversion and concern for the well-being of others, seem to be more deliberate and involve higher-order cognitive processes. The authors "propose that the first automatic response to inequality is mainly self-related, whereas, concerns for the well-being of others are part of higher cognitive, deliberative or intuitive processes following the first automatic response" (Alexopoulos et al., 2012, p. 8).

In an fMRI study investigating empathy toward others, Singer et al. (2006) found that both male and female participants exhibited empathy toward fair players. However, these empathy-related responses were significantly reduced in males observing an unfair person receiving pain. This reaction was correlated with an activation in the reward circuitry of the brain, indicating a desire for revenge. The activation of the reward circuitry of the brain when observing unfair people receive pain gives credence to the old adage that "revenge is sweet." Feng, Luo, and Krueger (2015) suggest that fairness-related norm enforcement recruits an intuitive system for rapid evaluation of norm violations and a deliberate system for integrating both social norms and self-interest to regulate the intuitive system in favor of more flexible decision making.

Another reason that could explain altruistic punishment is inequity aversion. Haruno and Frith (2010) found that the degree of inequity aversion in people described as prosocial is predicted by amygdala activity and unaffected by cognitive load. Based on these findings, they suggest that automatic emotional processing in the amygdala lies at the core of prosocial value orientation. Raihani and McAuliffe (2012) showed that participants punished cheats only when cheating produced disadvantageous inequity, while there was no evidence for reciprocity. The authors concluded that inequity aversion is an important determinant in punishing unfair behavior.

Hewig et al. (2011) showed that feedback-related negativity, an event-related brain potential, predicts the decision to reject unfair offers in the Ultimatum Game (UG). Feedback-related negativity (FRN) originates in the anterior cingulate cortex. The authors also found that the decision to reject unfair offers was also positively related to more negative emotional reactions and to increased nervous system activity. "Unfair offers in the one-shot UG were rejected more frequently, evoked more negative subjective emotional ratings, led to greater skin conductance responses (SCRs), and elicited larger feedback negativity (FNs), than fair offers" (Hewig et al., 2011, p. 512). The amygdala controls skin conductance responses through the regulation of the sympathetic nervous system. "Positive somatic markers associated with an action increase the likelihood of its selection, whereas negative somatic markers of an action decrease the likelihood of its selection" (Hewig et al., 2011, p. 513). Thus, affective somatic markers may contribute to the rejection of unfair offers.

Fliessbach et al. (2012) studied the neural basis of reactions to advantageous inequity (AI) and disadvantageous inequity (DI). Advantageous inequity is an unfair situation that benefits the individual, whereas disadvantageous inequity is an unfair situation that is detrimental to the individual. The authors showed hypoactivation of the ventral striatum under DI but not under AI; inequity-induced activation of the right dorsolateral prefrontal cortex was stronger under DI than AI; and correlations between subjective evaluations of AI evaluation and bilateral prefrontal and left insular activity. Any form of inequity represents a norm violation. Lord et al. (2010) contend that self-regulation includes both automatic and deliberate processes. Automatic processes are fast and unconscious, whereas deliberate processes are more elaborate and conscious.

Altruistic punishment could be the result of an emotional response to unfairness (Crockett et al., 2008). Crockett et al. (2010) note that "responders who reject a high proportion of unfair offers in the UG do so because they more strongly discount the delayed monetary benefits of accepting the offers relative to the immediate satisfaction of rejecting the offers" (p. 859). Consequently, the authors conclude that "altruistic punishment is an impulsive emotional reaction to perceived unfairness rather than a deliberative, goal-directed process" (ibid.). In another study, Crockett et al. (2010) showed that serotonin alters moral judgment and behavior through increasing aversion to personally harming others.

In economic exchange tasks, fairness refers to the equitable distribution of an initial stake of money between two people (Tabibnia et al., 2008, p. 339). This assumption suggests that fairness in economics is often defined in terms of distributive justice. However, for organizational justice scholars, fairness encompasses three dimensions: distributive justice, procedural justice, and interactional justice (Greenberg, 1987, 1990). Distributive justice refers to the fairness of outcomes received by an individual, whereas procedural justice refers to the fairness of the procedures underlining the distribution of outcomes. Interactional justice refers to the treatment of others with respect and dignity (Bies and Moag,

1986). Social comparison facilitates recipients' attention to the norm of fairness (Bohnet and Zeckhauser, 2004). To the extent that several studies in social cognitive neuroscience (Lieberman, 2007a, 2007b; Tabibnia et al., 2008) and neuroeconomics (Sanfey et al, 2003; Camerer et al., 2005) acknowledged the neural basis of fairness, it is important for organizational justice scholars to determine whether such studies could provide insights into understanding theories of organizational justice.

3 IMPLICATIONS FOR ORGANIZATIONAL JUSTICE THEORIES

As outlined above, neuroscientific research on fairness (and unfairness) can have implications for organizational justice scholarship. This research shows that people are sensitive to issues of fairness and reactions to fairness and unfairness have neural underpinnings. Brain structures, such as the anterior insula (AI), the nucleus accumbens, the orbitofrontal cortex (OFC), the anterior cingulate cortex (ACC), the lateral prefrontal cortex (lPFC), and the superior temporal sulcus (STS) are activated when people experience fairness. However, brain regions activated when people experience unfairness include the anterior cingulate cortex (ACC), the insula, the dorsolateral prefrontal cortex (dlPFC), the dorsomedial prefrontal cortex (dmPFC), the medial prefrontal cortex (mPFC), the superior temporal sulcus (STS), the ventral striatum (VS) and the ventrolateral prefrontal cortex (vlPFC) (see Table 7.1).

Neuroscientific research of fairness also shows that people tend to punish those who act unfairly toward others – the construct of altruistic punishment. The neuroscientific findings on reactions to fairness and unfairness and altruistic punishment could provide insights to organizational justice literature. Although organizational scholars have analyzed employee reactions to issues of fairness in the workplace, they have rarely explored its neural underpinnings. Recently, however, some organizational justice scholars have attempted to analyze the neural foundations of fairness in the workplace (Beugré, 2009; Dulebohn et al., 2009, 2016). Beugré (2009) developed a conceptual model of neuro-organizational justice that advocates that knowledge gleaned from the study of fairness in neuroeconomics and social cognitive neuroscience could illuminate our understanding of organizational justice. Dulebohn et al. (2009) provided the first neuroscientific evidence for two dimensions of organizational justice: distributive and procedural justice. Using fMRI, they found that unfair procedures evoked greater activation in parts of the brain related to social cognition, such as the ventrolateral prefrontal cortex and the superior temporal sulcus, whereas unfair outcomes evoked greater activation in more emotional areas of the brain, such as the anterior cingulate cortex, the anterior insula, and the dorsolateral prefrontal cortex.

The neural studies of fairness can help advance the organizational justice literature. In the following section, I explore the extent to which a neuroscientific approach could help elucidate our understanding of several organizational justice concepts, such as distributive justice, procedural justice, interactional justice and the major theories of deontic justice (Folger, 1998, 2001), fairness heuristics (Lind and Van den Bos, 2002; Van den Bos and Lind, 2002, 2004), and uncertainty management (Lind and Van den Bos, 2002; Van den Bos and Lind, 2002).

3.1 Neuroscience and the Three Dimensions of Organizational Justice

As indicated elsewhere in this chapter, organizational justice scholars often identify three main dimensions of organizational justice: distributive justice, procedural justice, and interactional justice (Greenberg, 1987, 1990). To what extent can knowledge gleaned from research in neuroscience help to improve our understanding of these three dimensions?

Distributive justice deals with how fair outcomes are allocated between recipients. According to Deutsch (1975, 1985), there are three ways to allocate outcomes: equity, equality, and need. The equity rule emphasizes the extent to which outcomes are proportionally distributed, whereas the equality rule emphasizes providing the same to everyone. The need-based rule implies that the outcomes should be distributed according to recipients' needs.

Most research on the neural basis of fairness tends to equate fairness to outcome distribution, that is, distributive justice (Dulebohn et al., 2009). This research mostly uses the Ultimatum Game. The main conclusions are that people reject unfair offers and there is neural evidence to support such rejections. Several brain regions are activated when participants reject unfair offers. Extended to organizational justice research, Dulebohn et al. (2009) showed that unfair outcomes evoked greater activation in more emotional areas of the brain, such as the anterior cingulate, anterior insula, and the dorsolateral prefrontal cortex. These findings could explain the extent to which employees are often sensitive to issues of pay equity.

A particular form of distributive justice is equity theory (Adams, 1965). People feel a sense of equity when they observe that their input/output ratio is equal to that of a comparison other. However, when the ratio is unequal and disadvantageous to the focal individual, he or she experiences a sense of disadvantageous inequity. This disadvantageous inequity may lead to inequity-reduction behaviors, such as reducing one's inputs, asking for an increase of outputs, changing the comparison other or quitting. The sense of disadvantageous inequity may arouse strong emotions. This emotional reaction is observed in the Ultimatum Game when participants reject unfair offers. In the case of advantageous inequity, people often tend to rationalize the situation to reduce their sense of guilt.

Several scholars have attempted to assess the neural basis of equity and inequity. For example, Cappelen et al. (2014) found a significant hemodynamic response in the striatum to deviations from the distribution of income that was proportional to work effort, but found no effect of deviations from the equal distribution of income. They also observed a striking correlation between the hemodynamic response in the striatum and the self-reported evaluation of the income distributions. Hsu et al. (2008) combined distribution choices with fMRI to investigate the central problem of distributive justice: the trade-off between equity and efficiency. They found that the putamen responds to efficiency, whereas the insula encodes inequity, and the caudate/septal subgenual region encodes a unified measure of efficiency and inequity (utility). They also found that individual differences in inequity aversion correlate with activity in inequity and utility regions.

Research has also looked at the neural basis of procedural justice, defined as the fairness of formal procedures that underline outcome distribution (Greenberg, 1987, 1990). According to Leventhal (1980), procedures should be bias free, consistently applied, accurate, correctable, representative of all, and ethical. Although most neuroscientific studies of fairness have focused on outcome distribution rather than the fairness of the processes, Dulebohn and colleagues conducted two series of studies that explored the neural basis of procedural justice. In the first study (Dulebohn et al., 2009), Dulebohn and his colleagues demonstrated that unfair procedures evoked greater activation in parts of the brain related to social cognitions, such as the ventrolateral prefrontal cortex and the superior temporal sulcus. In the second study (Dulebohn et al., 2016), they found that activation in the ventromedial prefrontal cortex and ventral striatum brain regions during procedural justice evaluation was associated with offer rejection in females, but not in males.

An organizational justice dimension that has not received research attention in terms of its neural underpinnings is interactional justice. Interactional justice refers to the fairness of interpersonal treatment, and generally deals with treating others with respect and dignity. For example, in organizations, employees expect their managers to show them respect. Interpersonal mistreatment might elicit strong emotional relations that could activate the emotional circuitry of the brain. However, experimental research in neuroscience has yet to address this issue. Organizational justice scholars who wish to analyze the neural basis of fairness in the workplace could conduct experiments involving participants' reactions to interpersonal mistreatment.

In addition to these three justice dimensions, neuroscientific research into fairness could enrich theories of organizational justice.

3.2 Neuroscience and Deonance Theory

Deonance theory construes justice as a moral obligation and as such advocates that people should act fairly because it is the right thing to do (Folger, 2001). In deontic justice, fairness is an end in itself. Indeed, people act fairly because they possess personal standards of justice and feel a moral duty to uphold these standards (Folger and Glerum, 2015). Recently, Cropanzano, Massaro, and Becker (2017) proposed a model of deontic justice with three dimensions: (1) use of justice rules to assess events; (2) cognitive empathy; and (3) affective empathy. A justice rule is a "self-based standard, or expectation, derived from individuals' socialized or internalized values, regarding the moral obligations of individuals in a specific context" (Lau and Wong, 2009, p. 281). The three dimensions of distributive justice, procedural justice and interactional justice are considered justice rules. Justice rules are personal standards of what is fair or not. To some extent they are yardsticks against which to evaluate an act as fair. The development of such standards could be rooted in context, such as the socialization process and the cultural environment in which people live (Cropanzano et al., 2017).

Cognitive empathy involves knowing people's feelings through deliberate thought, whereas affective empathy involves automatic sharing in the emotional experiences of others (e.g., Walter, 2012). According to Zaki and Ochsner (2012), cognitive and affective empathy reinforce each other. Cropanzano et al. (in 2017) also assert that people apply deontic justice principles because they show empathy for others. Deontic emotional experiences are derived from individuals' moral assumptions regarding how human beings should be treated and motivate third parties to punish rule-breakers (Folger, 2001).

Neuroscientific evidence shows that the three elements of deontic justice discussed by Cropanzano and colleagues have neural underpinnings. In fact, there is neuroscientific evidence that moral judgments have neural correlates (Greene et al., 2001; Moll et al., 2002; Prehn et al.,

2008). Moll et al. (2002) demonstrate that viewing moral and non-moral pleasant visual stimuli activates a common network of brain areas that includes the amygdala, the orbitofrontal cortex (OFC), the medial frontal gyrus (MedFG), and the cortex surrounding the right posterior superior temporal sulcus (pSTS). Moll, Oliveira-Souza, and Zahn (2008) noted that moral cognitions had a neural basis and identified brain regions such as the dorsolateral prefrontal cortex (dlPFC), the anterior prefrontal cortex (aPFC), the medial orbitofrontal cortex (mOFC), the ventromedial prefrontal cortex (vmPFC), the amygdala, and the STS region as implicated in moral cognition and behavior. These findings indicate that deontic justice as defined by organizational justice scholars could have a neural basis. Thus, framing justice in moral terms may activate these brain regions. Neuroscientific studies of fairness could also help to refine fairness heuristics theory.

3.3 Neuroscience and Fairness Heuristics Theory

Research in cognitive neuroscience indicates that System 1 and System 2 could be implicated in fairness judgments (Lieberman et al., 2002; Kahneman, 2003, 2011; Satpute and Lieberman, 2006). System 1 is automatic, intuitive, and effortless, whereas System 2 is deliberative, conscious, and effortful. These systems could help to explain some of the assumptions of fairness heuristics theory. Fairness heuristics theory posits two basic premises. First, fairness judgments are assumed to serve as a proxy for interpersonal trust in guiding decisions about whether to behave in a cooperative fashion in social institutions. Second, people are assumed to use a variety of cognitive shortcuts to ensure that they have a fairness judgment available when they need to make decisions about engaging in cooperative behavior (Lind, 2001).

Wu, Zhou et al. (2011) suggest that brain activity in evaluating fairness in division of assets entails both an earlier (semi-) automatic process in which the brain responds to fairness at an abstract level, and a later appraisal process in which factors related to social comparison and fairness norms come into play. Such fairness judgments using heuristics could involve System 1, which is automatic and effortless. However, fairness judgments using deliberate actions could involve System 2. Another element that could be relevant to the application of neuroscience to fairness heuristics is reputation. Nowak et al. (2000) note that the evolution of fairness, like the evolution of cooperation is linked to reputation.

In the Ultimatum Game, "accepting low offers damages the individual's reputation within the group and increases the chance of receiving reduced offers in subsequent encounters. Rejecting low offers is costly, but the cost is offset by gaining the reputation of somebody who insists on a fair offer" (Nowak et al., 2000, p. 1774). It is worth noting that "humans are accustomed to repeated interactions" (Nowak et al., 2000, p. 1773). It is obvious that organizations are arenas of such repeated interactions. Hence, reputation influences interactions between individuals. People who have a good reputation may command respect and admiration from their peers. However, those who have a bad reputation may incur negative reactions such as avoidance or even mistreatment. Hence, when people lack information about others, they may rely on others' reputation to decide whether to act fairly or unfairly toward them. Lacking information about others may create a situation of uncertainty, which is addressed by uncertainty management theory.

3.4 Neuroscience and Uncertainty Management Theory

Van den Bos and Lind (2002) introduced uncertainty management theory as a framework to explain why people care about fairness. According to the authors, people use fairness as a means of addressing uncertainty. Berger (1979) identified two types of uncertainty: cognitive uncertainty – the degree of uncertainty associated with attitudes and beliefs – and behavioral uncertainty – the extent to which behavior is predictable. Social interactions entail some form of uncertainty. Hence, when people interact with others, they try to reduce this uncertainty by finding ways to predict others' behavior. One of the best ways to reduce uncertainty in social interactions is to gather information about the other party. At the interpersonal level, uncertainty occurs either when a person confronts an incompatibility between different cognitions, between cognitions and experiences, or between cognitions and behavior (Van den Bos and Lind, 2002). Although uncertainty may be construed as inherent to human nature, people have a managerial as well as psychological need to feel certain about their business environment (ibid.). Van den Bos and Lind (2002) argued that people attempting to eliminate uncertainty might also try to find some way to make it tolerable and cognitively manageable.

Uncertainty management theory uses two principles – the substitutability principle and the primacy effect – to explain why people care about justice and how they form fairness judgments. The substitutability principle contends that, because information about outcomes is not often available but information about procedures is, people rely on the latter to make judgments of fairness concerning their outcomes (Van den Bos and Lind, 2002). In other words, people substitute information about procedures for information about outcomes. Thus, receiving fair outcomes may imply that the processes underlying this outcome distribution were fair. The primacy effect implies that information that comes first exerts a stronger influence on fairness judgments than information that comes second. Consequently, the primacy effect contends that information that is first used to make judgments of fairness tends to influence subsequent fairness judgments. Neuroscience research on uncertainty could provide insight for scholars studying uncertainty management theory – previous research has indicated activation of different brain regions, including the amygdala and the insula (Camerer et al., 2004; Vartanian et al., 2011; Ma and Jazayeri, 2014).

8. The neural basis of trust and cooperation

Humans are biologically hardwired to engage in exchange, which requires trust and cooperation. Although this chapter focuses on the neural basis of trust and cooperation, the two concepts are different. Trust refers to a person's "expectations, assumptions, or beliefs about the likelihood that another's future actions will be beneficial, favorable, or at least not detrimental to one's interests" (Robinson, 1996, p. 576). Defined as such, trust can be construed as a probability. Suppose two people, A and B, are involved in some form of interaction. By trusting person B, person A is assuming that person B will not engage in actions that would be detrimental to him or her. In doing so, person A is accepting being vulnerable to the actions of person B. Hence, trust entails a sense of vulnerability. In fact, trust implies a willingness to be vulnerable.

"Trust is indispensable in friendship, love, families, and organizations and it is a lubricant of economic, political, and social exchange" (Baumgartner et al., 2008, p. 646). "Individuals' judgments about others' trustworthiness are anchored, at least in part, on their a priori expectations about others' behavior … and those expectations change in response to the extent to which subsequent experience either validates or discredits them" (Kramer, 1999, p. 576). Trust reduces transaction costs and enhances interpersonal relations because trusting parties tend not to safeguard themselves (Granovetter, 1985; Williamson, 1993; Bromiley and Cummings, 1996). As such, many relations in the workplace depend on trust and cooperation. It is understood that these two forms of social interactions are believed to have neural underpinnings, as demonstrated by research in neuroeconomics.

1 UNDERSTANDING TRUST AND COOPERATION IN SOCIAL EXCHANGE

Arrow (1974) argues that trust denotes cooperative conduct, whereas distrust denotes opportunistic behavior. Fukuyama (1995) construes trust as a form of social capital. As social capital, trust engenders spontaneous sociability among partners. In organizations, spontaneous sociability leads to organizational members' willingness to cooperate with others to perform specific tasks. In communities, spontaneous sociability translates into reciprocal behavior and cooperation. Cooperation refers to joint actions between parties for a particular purpose.

Benkler (2011a, 2011b) contends that humans are more cooperative than we ordinarily think and "the ability to trust is a key element in cooperation" (Benkler, 2011b, p. 8). Because people have a natural tendency to cooperate, Benkler (2011b) argues that "we need systems that rely on engagement, communication, and a sense of common purpose and identity" (p. 5) and suggests seven ways for building systems that foster cooperation among their agents: (1) communication; (2) framing and authenticity; (3) empathy and solidarity; (4) fairness and morality; (5) rewards and punishment; (6) reputation and reciprocity; and (7) diversity.

Nowak (2006) notes that "perhaps the most remarkable aspect of evolution is its ability to generate cooperation in a competitive world. Thus, we might add 'natural cooperation' as a third fundamental principle of evolution beside mutation and natural selection" (p. 1563). According to him, humans are not selfish and only cooperate more with those they share the same genes with. To the contrary, they are wired to cooperate with others. Indeed, "cooperation is the decisive organizing principle of human society" (p. 1560). These arguments are contrary to the contention of biologist Richard Dawkins (1976) who argues that human beings are born selfish.

Trust is required in the workplace for collaborative work. For example, equity theory (Adams, 1965) suggests that employees believe that their outcomes should be proportional to their inputs compared to relevant others. The occurrence of such events would lead to trust. The reciprocation of trust would be for employees to work hard and show commitment and loyalty to their organization. Whereas trust elicits reciprocation, distrust engenders non-reciprocation. Because it enhances reciprocity, trust plays a critical role in interpersonal cooperation. Other benefits of trust include reducing transaction costs, increasing spontaneous sociability, and facilitating appropriate forms of deference to organizational authorities (Kramer, 1999, p. 582).

Trust is an essential ingredient of social exchange and represents the glue that binds relationships together. Several scholars (Arrow, 1974; Kramer, 1999) have emphasized the role of trust in social exchange and interpersonal relationships. An interesting question, however, is whether people are better off trusting than not trusting. This is a tricky question because we have been taught that trust is essential in any exchange relationship between two parties. However, a close look at the trusting relationship reveals that it is more complicated than it first appears. Suppose that I make the argument that one is better off not trusting than trusting. The foundation of the argument is explained as follows.

Suppose that you are involved in a transaction with another party in which trust is important. Should you trust the other party or not? Using the combinatorial formula of the French philosopher and mathematician Blaise Pascal, whereby permutations of objects are combined to make sense of reality, would suggest that one is better off not trusting than trusting. And you may be right for not trusting the other party. If you trust and the other party is trustworthy and acts accordingly, you end up being better off. However, if you trust and the other party is deceptive, then you end up being worse off. The probability of being better off when trusting is only 0.5. Suppose now that you do not trust the other party and the other party is deceptive, you are better off because you would have safeguarded yourself against any mischief. Suppose also that you do not trust the other party but the other party is indeed trustworthy and acts accordingly, you are also better off. The probability of being better off when not trusting is 1 (Table 8.1).

Table 8.1   Trust/distrust scenario

	Trust	Distrust
Trustworthy	Better off	Better off
Untrustworthy	Worse off	Better off
Probability of being better off	0.5	1

The scenario described above would suggest that people should distrust rather than trust others. Since the probability of being better off when distrusting ($p = 1$) is greater than the probability of trusting ($p = 0.5$), this would make perfect sense. However, trust and distrust bear outcomes. Trust reduces transaction costs, whereas distrust increases them. Take the example of a new mother who hires a nanny to take care of her three-month-old baby. The mother may decide to trust or distrust the nanny. If she trusts the nanny, she may not take any monitoring action, thereby lowering the costs of this transaction. If the nanny takes care of the baby as expected, the mother is better off (and the baby also). However, if the nanny fails to properly care for the baby, the mother is worse off (and the baby also). Suppose now that the mother does not totally trust the nanny and as a consequence monitors her behavior by placing a camera in her house. This monitoring device may deter the nanny from engaging in deceptive behaviors. In this case, the mother and the baby are better off. However, distrust would increase the costs of the transaction because the monitoring device is costly. Although trust is important in human relations, it is important to acknowledge that trust is not a gift to give but a value to earn.

Montague, Lohrenz, and Dayan (2015) singled out three mechanisms of trust: reaping, regarding, and recursive modeling. Reaping involves mechanisms that respond to punishment or reward. Regarding comprises other-regarding or prosocial mechanisms. Recursive modeling encompasses hierarchical cognitive modeling of others through exchanges. Trust can be based on past interactions with others. For example, if a person has acted in a way we expected in the past, we are likely to use this experience as an anchor for future interactions with the person. Trust can therefore act as a heuristic device. Indeed, "individuals' judgments about others' trustworthiness are anchored, at least in part, on their a priori expectations about others' behavior ... and those expectations change in response to the extent to which subsequent experience either validates or discredits them" (Kramer, 1999, p. 576).

2 NEURAL STUDIES OF TRUST

Biological investigations into trust have been carried out on three levels of analysis: genes, endocrinology, and the brain. Riedl and Javor (2012) reviewed the literature on the biology of trust and found that trust behavior is at least moderately genetically predetermined, and is associated with specific hormones, in particular oxytocin, as well as specific brain structures, which are located in the basal ganglia, limbic system, and the frontal cortex. The Trust Game developed by Berg, Dickhaut, and McCabe (1995) is often used to study the neural basis of trust and cooperation. In this game, two players are randomly and anonymously matched, one as investor and the other as trustee. The amount sent by the investor is considered as a measure of trust and the amount returned by the trustee is a measure of trustworthiness.

In studying the neural basis of trust, researchers are mostly concerned about the role of certain brain structures on trusting behavior. King-Casas et al. (2005) conducted a multiround Trust Game on a sample of 48 participants and found that the intention to trust was coded in the caudate nucleus. In fact, the head of the caudate nucleus receives or computes information about (1) the fairness of a social partner's decision; and the (2) intention to repay that decision with trust. Delgado, Frank, and Phelps

(2005) had participants play the Trust Game with three hypothetical partners depicted as being good, bad, or neutral. They found that participants choose to be more cooperative with the good partner. They also found that participants were persistently more likely to make risky choices with the "good" partner than with the other types of partners. Activation in the caudate nucleus differentiated between positive and negative feedback, but only for the "neutral" partner. It did not do so for the "good" partner and did so only weakly for the "bad" partner, suggesting that prior social and moral perceptions can diminish reliance on feedback mechanisms in the neural circuitry of trial-and-error reward learning. The authors concluded that "activation of the ventral striatum may indicate that participants possibly experienced the presentation of the good partner, or the idea of transferring funds to him or her, as rewarding in itself (Delgado et al., 2005, pp. 1615–1616).

Winston et al. (2002) determined the neural basis of trustworthiness judgments using event-related functional magnetic resonance imaging (fMRI). They found that trustworthiness was correlated with blood oxygenation level-dependent (BOLD) signal change to reveal task-independent increased activity in the bilateral amygdala and right insula in response to faces judged to be untrustworthy. Right superior temporal sulcus showed enhanced signal change during explicit trustworthiness judgments alone. Fan et al. (2011) explore the existence of a potential neural network for empathy. Using a whole brain-based quantitative meta-analysis of recent fMRI studies of empathy, Fan et al. (2011) found that the dorsal anterior cingulate cortex–anterior mid-cingulate cortex–supplementary motor area (dACC–aMCC–SMA) and bilateral anterior insula were consistently activated in empathy. The dorsal aMCC was activated in the cognitive evaluation form of empathy, whereas the right anterior insula was involved in the affective-perceptual form of empathy only. The left anterior insula was active in both forms of empathy. In this study, the authors observed that affective-perceptual empathy occurs automatically through observation, whereas cognitive-evaluative empathy occurs when the person attends to the feelings of the target.

In a study of 82 healthy participants, Haas et al. (2015) found that participants characterized as trusting others exhibited increased gray matter volume within the bilateral ventromedial prefrontal cortex and bilateral anterior insula than those that were not so characterized. Greater right amygdala volume is associated with the tendency to rate faces as more trustworthy and untrustworthy. A whole brain analysis also showed that the tendency to trust was reflected in the structure of the dorsomedial prefrontal cortex.

Wardle et al. (2013) used an event-related fMRI design to examine what neural signals correspond to experimentally manipulated reputations acquired in direct interactions during trust decisions. They found that the caudate (both left and right) signals reputation during trust decisions, such that it is more active to partners with two types of "bad" reputations, either indifferent partners (who reciprocate 50 percent of the time) or unfair partners (who reciprocate 25 percent of the time), than to those with "good" reputations (who reciprocate 75 percent of the time). These findings show that the caudate is involved in signaling and integrating reputations gained through experience into trust decisions, demonstrating a neural basis for this key social process. The caudate signals the presence of bad or risky partners.

Hahn et al. (2015) show that a person's initial level of trust is, at least in part, determined by brain electrical activity acquired well before the beginning of an exchange. The initial level of trust is important in relations with strangers. Even if a person does not have past experiences with another person, he or she could still trust this person when interacting with him or her. This initial level of trust is updated in further exchanges with the party. Hence, "reciprocal exchanges can be understood as the updating of an initial belief about a partner" (Hahn et al., 2015, p. 809). Indeed, cooperation between people cannot occur without a minimum level of trust. Therefore, we can speculate that initial trust is important in starting a relationship. Such initial trust is not the product of previous exchanges with the same person.

Fett et al. (2014) found that increasing age was associated with higher trust at the onset of social interactions, increased levels of trust during interactions with a trustworthy partner, and a stronger decline in trust during interactions with an unfair partner. Indeed, people become more trusting of others as they age. The authors also found that increased brain activation in mentalizing regions, such as the temporoparietal junction, the posterior cingulate and the precuneus, supported this behavioral change. Age was associated with reduced activation in the reward-related orbitofrontal cortex and caudate nucleus during interactions with a trustworthy partner, possibly reflecting stronger expectations of trustworthiness.

2.1 The Role of Oxytocin in Trusting Behavior

The neurotransmitter oxytocin (OT) has been linked to enhanced trust and cooperation. "OT is a neuropeptide hormone that is synthesized in the hypothalamus and is released in both the brain and the periphery" (Nave, Camerer, and McCullough, 2015, p. 772). OT has the potential to not only modulate activity in a set of specific brain regions, but also the functional connectivity between these regions (Bethlehem et al., 2013). Carter (2014) discusses the role of oxytocin in human behavior and concludes that oxytocin acts to enable the high levels of social sensitivity and attunement necessary for human sociality and for rearing a human child. She also argues that under optimal conditions, oxytocin may create an emotional sense of safety. Kosfeld et al. (2005) showed that intranasal administration of oxytocin increases the levels of trust among participants, thereby facilitating social interactions. They also showed that the effect of oxytocin on trust is not due to a generalized increase in the readiness to bear risks. Rather, oxytocin affects an individual's willingness to accept social risks arising through interpersonal interactions.

According to De Breu and Cret (2016), oxytocin motivates and enhances the ability to (1) like and empathize with others in their groups; (2) comply with group norms and cultural practices; and (3) extend and reciprocate trust and cooperation. Oxytocin also motivates intergroup discrimination and emphasizes intragroup love and acceptance.

Baumgartner et al. (2008) examined the neural circuitry of trusting behavior by combining the intranasal, double-blind administration of oxytocin with fMRI. They found that participants in the oxytocin group showed no change in their trusting behavior after learning that their trust had been violated several times while those receiving placebo decrease their trust. The authors also found that this difference in trust adaptation was associated with a specific reduction in activation in the amygdala, the midbrain regions, and the dorsal striatum in participants who had received oxytocin. They interpreted these findings by suggesting that the neural systems mediating fear processing (the amygdala and the midbrain regions) and behavioral adaptations to feedback information (dorsal striatum) modulate oxytocin's effect on trust.

Zak, Kurzban, and Matzner (2005) studied the effect of oxytocin on trust in humans using the Trust Game. Trust and trustworthiness were measured using the sequential anonymous Trust Game with monetary payoffs. They found that oxytocin levels were higher in participants who received a monetary transfer that reflected an intention of trust relative to an unintentional monetary transfer of the same amount. They also found that higher levels of oxytocin were associated with trustworthy behavior (the reciprocation of trust). Zak et al. (2005) also found that in the absence of intentionality, both oxytocin and behavioral responses were extinct. They concluded that perceptions of intentions of trust affect levels of circulating oxytocin. Dimoka (2010) studied trust and distrust in the context of management information systems. She found that trust and distrust activated different regions of the brain. Trust was associated with the brain's reward circuitry, whereas distrust was associated with the brain's intense emotions and fear of loss areas. Particularly, trust was associated with the caudate nucleus and putamen that constitute the dorsal striatum, whereas distrust was associated with the amygdala.

McCabe et al. (2001) found that players who cooperated more often with others showed increased activation in Brodmann area 10 (BA10) and in the thalamus. Feng et al. (2015) studied the role of OT and vasopressin on social behavior. In a double-blind, placebo-controlled study, 153 men and 151 women were randomized to receive 24 IU intranasal OT, 20 IU intranasal arginine vasopressin (AVP) or placebo. The participants received fMRI while playing an altered version of the Prisoner's Dilemma Game with same-sex partners. They observed that OT increased the caudate/putamen response among males, whereas it decreased this response among females. Similar sex differences were also observed for AVP effects within the bilateral insula and right supra-marginal gyrus when a more liberal statistical threshold was employed. OT increased the BOLD fMRI response to reciprocated cooperation from human partners in males, whereas OT decreased the BOLD fMRI response to reciprocated cooperation in females.

2.1.1 Oxytocin and in-group/out-group trust and cooperation

Despite the apparent link between oxytocin and trust behavior, several studies have indicated that the effect of oxytocin tends to be limited to in-group members and oxytocin could even lead to aggression toward out-group members, specifically when this group is perceived as threatening. Chen, Kumsta, and Heinrichs (2011) contend that the effect of oxytocin on trust and cooperation seems to be exercised more for in-group members than out-group members. Thus, they conclude that trust and goodwill-inducing effects of oxytocin are relatively limited. De Breu et al. (2011a) contend that oxytocin tends to facilitate ethnocentrism. It leads to trust and cooperation among in-group members but induces bias and non-cooperation with out-group members. In a set of experiments, the authors found that oxytocin creates intergroup bias because it motivates in-group favoritism and to a lesser extent, out-group derogation.

The findings indicate that oxytocin's effect on trust and cooperation depends on the characteristics of the target. Oxytocin tends to create positive feelings about similar others and derogation toward dissimilar others. "In-group favoritism has strong adaptive value and facilitates within group coordination and survival" (De Breu et al., 2011a, p. 1264). "Oxytocin creates intergroup bias primarily because it motivates in-group favoritism and not because it motivates out-group derogation" (De Breu et al., 2011a, p. 1265). In a subsequent study, De Breu et al. (2011b) conclude that oxytocin-induced goodwill tends to be limited only to in-group members. These studies indicate that oxytocin-induced trust and cooperation are not directed toward everyone. Oxytocin-induced goodwill (empathy, trust, and cooperation) is contingent upon the perceived characteristics of the targets. "The oxytocin circuitry may have evolved to sustain within-group cooperation, in-group protection, and if needed, competition towards rivaling out-groups" (De Breu, 2012, pp. 424–425).

In an extensive literature review on the effect of oxytocin on human behavior, De Breu notes that oxytocin modulates the regulation of cooperation and conflict among humans for three reasons: (1) oxytocin enables social categorization of others into in-group and out-group; (2) oxytocin dampens amygdala activity and enables the development of trust; and (3) oxytocin up-regulates neural circuitries, such as the inferior frontal gyrus, the ventromedial prefrontal cortex, and the caudate nucleus, and is involved in empathy and concern for others. Because "cooperation serves individual and group survival in ancestral and contemporary societies, it may have its root cause in evolved neuro-biological circuitries" (De Breu, 2012, p. 420). From an evolutionary standpoint, oxytocin motivates in-group favoritism, cooperation toward in-group but not out-group members, and defense-motivated non-cooperation toward threatening outsiders (De Breu, 2012). De Breu (2012, p. 419) argues that oxytocin's primary functions include in-group "tend-and-defend."

Despite the positive effect of oxytocin on trust and cooperation, several authors suggest that the literature on the relationship between oxytocin and trust should be considered with caution. Kanat, Heinrichs, and Domes (2014) reviewed the literature exploring the link between OT and social behavior and concluded that there is still a general lack of basic knowledge regarding the distribution of OT receptors and the molecular and cellular mechanisms of OT in the human brain. Hence, findings should be treated with caution and it is important for researchers to define the cognitive processes and social behaviors they are investigating. After an extensive review of the literature on the role of oxytocin in trusting behavior, Nave et al. (2015) conclude that the cumulative evidence does not provide robust convergent evidence that human trust is reliably associated with OT (or caused by it). Bartz et al. (2011) noted that a main effect of OT on target behavior was found in only half of the published studies. In addition, the effect of OT on prosocial behavior was weak.

3 NEURAL BASIS OF COOPERATION

Cooperation is essential for the functioning of human societies and is a fundamental characteristic of human kind. As Crockett, Clark, Lieberman, et al. (2010, p. 855) put it, "unlike most other species, humans cooperate in large groups, with strangers they are unlikely to encounter again, and often in the absence of immediate external reinforcement." The authors note that many of the processes underlying cooperation overlap with rather fundamental brain mechanisms, such as those involved in reward, punishment, and learning. Cooperation is highly associated with activation in brain areas involved in reward-based learning. "People learn about the cooperative nature of another player based on the history of that partner's behavior" (Stallen and Sanfey, 2013, p. 298). Anticipation of guilt may motivate people to cooperate as well as expectations of trustworthiness and reputation for being cooperative and trustworthy (Stallen and Sanfey, 2013).

3.1 Reciprocal Cooperation

Because of the ubiquitous nature of cooperation between human beings, one must address the following questions. Under which conditions do humans cooperate? Do they cooperate in situations of reciprocity when they know that the other party will cooperate? Cooperation in human interactions can be induced through reward or punishment. Indeed, people learn the value of reward as well as punishment. Incentives are quite effective in promoting cooperation. As indicated in Chapter 7, fear of punishment, including altruistic punishment, acts as a deterrent to free riding. People learn to trust others who have been trustworthy in the past. This probably explains the adage that trust is earned not given. People are also more likely to cooperate with in-group than out-group members.

Rilling, Sanfey, and Aronson (2004a) conducted an fMRI study on a sample of 19 participants and found that the ventromedial prefrontal cortex (vmPFC) and the ventral striatum were activated in the case of reciprocated and unreciprocated cooperation. The ventral striatum and the vmPFC showed an increased BOLD response to reciprocal altruism and a decreased BOLD response in unreciprocated altruism. In the experiment, participants played a single-shot Prisoner's Dilemma Game while their brains were being scanned. The authors conclude that "activity in the mesolimbic dopamine system may help us to learn whom we can and cannot trust to reciprocate favors, motivating us to seek out future interactions with the former and avoid future interactions with the latter"

(Rilling et al., 2004a, p. 5). These brain structures may help us to predict who we should cooperate with and avoid those we cannot.

Stevens and Stephens (2004) note that cooperation can occur when the individual's benefits depend completely on the actions of others. In humans, the concepts of reciprocity and altruistic reciprocity in which people exchange favors are considered important in inducing cooperation: the "you scratch my back, I scratch yours" principle. De Breu (2012) argues that social systems share two properties: (1) each individual member serves his or her personal interests best by opting for non-cooperation; and (2) when all members opt for non-cooperation, each is worse off than when all opt for cooperation.

Trivers (1971) proposes the concept of direct reciprocity as a mechanism for inducing cooperation between unrelated individuals. He argued that each individual has both altruistic and cheating tendencies. "Altruistic behavior can be defined as behavior that benefits another organism, not closely related, while being apparently detrimental to the organism performing the behavior, benefit and detriment being defined in terms of contribution to inclusive fitness" (Trivers, 1971, p. 35). Axelrod and Hamilton (1981) note that "while an individual can benefit from mutual cooperation, each one can also do even better by exploiting the cooperative efforts of others" (p. 1391). The authors used the Prisoner's Dilemma Game to illustrate their point. They note that "apart from being the solution in game theory, defection is also the solution in biological evolution" (ibid.). Social expectations influence people's behavior. For example, people tend to behave the way they think others expect them to behave. The right lateral prefrontal cortex appears to be involved in applying a pre-existing social norm to decision making (Sanfey, Stallen, and Chang, 2014).

Reputation, guilt, and shame all play a role in human cooperation (Tomasello and Vaish, 2013). One may cooperate or reciprocate cooperation with others to establish a reputation of being trustworthy and cooperative. One may also cooperate to avoid a sense of guilt for not reciprocating. One may also cooperate to avoid embarrassment. Social norms are widely shared sentiments about what constitutes appropriate behavior. Buckholtz and Marois (2012) note that strong reciprocity requires that individuals have the capacity to learn norms; integrate predictions about norm-related action outcomes into decision making to guide their own behavior; assess other individuals' beliefs, desires, and behavior in the context of these norms; and use subjective responses to norm violations to appropriately sanction defection.

The ventromedial prefrontal cortex (vmPFC) and ventral striatum are implicated in the generation of observational learning signals. Damage to the vmPFC diminishes cooperation and sense of guilt. The threat of punishment induces prosocial behavior, incentivizing cooperation in people who are otherwise inclined to defect. Buckholtz and Marois (2012, p. 659) reviewed the literature on cooperation and norms enforcement and conclude that "cooperation among unrelated individuals is predicated on the unique ability to establish norms, to transmit these norms from generation to generation and to enforce these norms through punishment." According to Bicchieri (2006), social norms represent the grammar of social interaction and serve to foster social peace, stabilize cooperation, and enhance prosperity.

3.2 Cooperation as a Rewarding Experience

Tabibnia and Lieberman (2007) argue that cooperation is rewarding and that some brain regions are activated when people cooperate with others. Non-monetary rewards also have "hedonic inputs to individuals' behaviors" (p. 90). According to the authors, positive feelings (and absence of negative feelings) seem to be associated not only with fair treatment and cooperation but also with those who are considered cooperative and trustworthy. Cooperation between people requires the ability to infer each other's mental states to form shared expectations. McCabe et al. (2001) had 12 participants play a standard "trust and reciprocity" game with both a human and a computer. They found that among the seven people attempting cooperation with one another or the machine, the prefrontal regions were more active when participants were playing a human than a computer. However, within the group of the five non-cooperators, there was no difference in prefrontal activation between computer and humans. "Cooperation requires an active convergence zone that binds joint attention to mutual gains with sufficient inhibition of immediate reward gratification to allow cooperative behavior" (McCabe et al., 2001, p. 11834).

Decety et al. (2004) argue that cooperation is a rewarding process. It is associated with the left medial orbitofrontal cortex. The authors found that the medial orbitofrontal cortex area was activated when people cooperate with another person. This is consistent with previous research suggesting that cooperation is more socially rewarding. Rilling, Gutman, and Zeh (2002) conducted an fMRI study on a sample of 36 women and found that mutual cooperation was associated with consistent activation in brain areas that have been linked with reward processing, such as the nucleus accumbens, the caudate nucleus, the ventromedial frontal cortex, the orbitofrontal cortex, and the rostral anterior cingulate cortex. Based on these findings, the authors propose that activation of this neural network positively reinforces reciprocal altruism.

Rilling, Sanfey, and Aronson (2004b) have participants play the Ultimatum Game and the Prisoner's Dilemma Game with both humans and a computer. Riolo, Cohen, and Axelrod (2001) found that although computer partners elicited activation of the same brain area activated by human partners, most of these activations were stronger for human partners. Indirect reciprocity applies when benevolence to one agent increases the chance of receiving help from others. Riolo et al. (2001) used computer simulations to show that cooperation can arise when agents donate to others who are sufficiently similar to themselves in some arbitrary characteristic. The authors contend that tag-based donations can lead to the emergence of cooperation among agents who have only rudimentary ability to detect environmental signals and, unlike models of direct or indirect reciprocity, no memory of past encounters is required. Tag refers to similarity between interactants and can facilitate cooperation without having memory of past encounters or information from third parties about the reputation of a potential beneficiary. In organizations, tag can represent artifacts such as cultural similarity (accent, same ethnicity, or gender).

3.3 Conditional Cooperation

Suzuki et al. (2011) studied the concept of conditional cooperation using fMRI. Conditional cooperation is cooperation based on the actions of another person. They found that participants cooperated more frequently with both cooperative and neutral opponents than with non-cooperative opponents. They also found that the right dorsolateral prefrontal cortex (dlPFC) was activated when participants confronted non-cooperative opponents. The dlPFC is involved in cognitive top-down inhibition or pre-potent responses. The authors also observe that participants who showed more activation in the parietal cingulate cortex were more likely to cooperate with others.

In general, people cooperate with others when those others have cooperated in previous interactions or have the reputation of being cooperative or would cooperate in the future. In this regard, conditional cooperation has elements of reciprocal cooperation and can even be perceived as a rational behavior. Rilling, Goldsmith and Glenn (2008) studied participants' reactions to unreciprocated cooperation using fMRI. They found that unreciprocated cooperation was associated with greater activity in the bilateral anterior insula, the left hippocampus, and the left gyrus, compared with reciprocal cooperation. These brain regions were also more responsive to unreciprocated cooperation than to successful risk taking in a non-social context. To some extent, these brain regions could be thought of as helping to express negative feelings related to unreciprocated cooperation.

Fehr and Fischbacher (2004) reviewed the literature on norms and social cooperation and conclude that: (1) sanctions are decisive for norm enforcement; and (2) they are largely driven by non-selfish motives. The existence of social norms creates conformity within groups and heterogeneity across groups. "Cooperation in human societies is based on social norms, including in modern societies, where a considerable amount of cooperation is due to the legal enforcement of rules" (Fehr and Fischbacher, 2004, p. 185). "Human cooperation is largely based on a social norm of conditional cooperation" (Fehr and Fischbacher, 2004, p. 186). "This norm prescribes cooperation if the other group members also cooperate, whereas the defection of others is a legitimate excuse for individual defection" (ibid.).

Take the example of teamwork in organizations. The contribution of an individual member could raise the output of the whole team and other team members. However, a defection by a team member can negatively impact the team and reduce the output of other individual members. This leads to what is generally labeled the "public good" problem. "Positive or negative side-effects of individual actions typically give rise to a cooperation or a 'public-good' problem" (Fehr and Fischbacher, 2004, p. 185). In the absence of sanctions, conditional cooperation may be reduced over time.

An interesting question is, why do people tend to be conditional cooperators? Why do people care about the average contribution of others before deciding to cooperate themselves? Chaudhuri (2011) reviews the extant literature on conditional cooperation and concludes that many participants are conditional cooperators whose contributions are positively correlated with their belief about the average group contribution. Conditional cooperators suffer monetary losses to punish free riders; they also use non-monetary mechanisms such as expressions of disapproval, advice giving, and assortative matching.

"Participants use information about the average group contributions as an anchor for their own future contributions" (Chaudhuri, 2011, p. 52). Because cooperation is costly, people may be more inclined to be conditional cooperators. They would only cooperate if doing so satisfies some expected outcome. However, if no one cooperates, everyone is worse off (Dawes and Messick, 2000).

Declerck, Boone, and Emonds (2013) propose that the motivation to cooperate (or not) is modulated by two neural networks: a cognitive control system and/or a social cognition system. The cognitive control system processes extrinsic cooperative incentives, whereas the social cognition system processes trust and other threat signals. These two drivers of cooperation are not exclusive. The authors speculate that self-regarding individuals are more responsive to external cooperative incentives, whereas other-regarding individuals are more sensitive to trust signals. Caceda et al. (2015) showed that intrinsic brain connectivity positively influences reciprocal social behavior. In human interactions, reciprocity is defined as "responding to a positive action by another positive action" (p. 479).

4   IMPLICATIONS FOR ORGANIZATIONS

Understanding the neural basis of trust and cooperation has implications for organizational phenomena such as team dynamics, team cohesiveness and effectiveness, organizational commitment, employee engagement and emotional contagion to name but a few. Without trust, organizational members cannot commit their life, effort, and career to an employer. Although the study of the role of trust and cooperation in organizations has a relatively long history in management and organizational behavior (McAllister, 1995; Kramer and Tyler, 1996; Lewicki, McAllister, and Bies, 1998), exploring its neural underpinnings is relatively new.

4.1   Implications for Teamwork

Organizational scholars have underscored the role of trust in team cohesion and effectiveness, transactive memory, and shared mental models (Kramer and Tyler, 1996; Waldman et al., 2015). Hence, knowing the neural foundations of trust could provide opportunities to organizations to create awareness on the inherent nature of trust. Such understanding could also allow organizations to understand how emotions, positive or negative, could be spread in teams. When employees trust each other and cooperate effectively, the mirror neuron system (Iacoboni, 2009) may provide an opportunity to consciously or unconsciously share several attributes. For example, the mirror neuron system may facilitate emotional contagion between team members. Positive as well as negative emotions may spread within the team through the mirror neuron system.

Another application of the neuroscience of trust to team dynamics is whether teams can be studied in a more naturalistic way. Current research on teamwork and team dynamics relies on survey methodologies used after the teams have been dismantled and performed the tasks for which they were created. A neuroscientific approach could help study teams when these teams are currently performing their tasks. Waldman et al. (2015) cite the example of Advanced Brain Monitoring (ABM), a California-based company that has developed quantitative electroencephalography (qEEG) software to study the neural patterns in human interactions. This equipment could help assess team neurodynamics in real time.

The neuroscience of trust could also help to improve transactive memory, and shared mental models (Waldman et al., 2015). Shared mental models lead to team performance (Klimoski and Mohammed, 1994; Mathieu et al., 2000; Mohammed and Dumville, 2001). The concept of shared mental models refers to an organized understanding of relevant knowledge that is shared by team members (Klimoski and Mohammed, 1994). Hence, when trust is high, team members may be more likely to share the goals and values of the team, thereby increasing team effectiveness. In addition to shared mental models, neuroscience could help minimize interteam competition and create an *esprit de corps*.

4.2 Fostering Cooperation in the Workplace Using Neuroscience Techniques

The neuroscience of cooperation has implications for cooperation in organizations. Although people perform individual tasks at work, organizations are arenas where human cooperation is the most needed. Hence, neuroscience could help managers build a strong case for cooperating behavior in the workplace. How can cooperation be enhanced in organizations using neuroscience findings? Should organizations administer oxytocin to their members to increase trust and cooperation? Doing so could raise serious ethical issues. However, training employees on the functioning of the human brain (and therefore their own brain) could create awareness, which could perhaps translate into behavior modification. Such cooperation could improve team effectiveness and strengthen organizations.

Cooperation is necessary for the survival of the human species. As Tabibnia and Lieberman (2007, p. 90) put it, "we live in a highly social environment, in which most of the work we do is accomplished through collaboration with others and many of the goods we consume are consumed in the company of others or shared with others." Cooperation is important for organizational life. Without cooperation, humans cannot build competitive and sustainable organizations and prosperous communities.

9. The neural basis of ethics and morality

Ethics and morality play an important role in human life. From ancient philosophers, including Aristotle, Plato, and Socrates, to modern economists, philosophers, psychologists, and sociologists, the problems of ethics and morality have been widely discussed. Although the study of ethics is an ancient tradition rooted in religious, cultural, and philosophical beliefs, its extension to business settings known as business ethics is of recent interest (Phillip, 1985; Trevino and Brown, 2004). Today, organizational scholars consider ethics important to management education and practice (Christensen et al., 2007), particularly in the aftermath of numerous corporate scandals, including Enron in the 1990s and Volkswagen in the 2010s.

Christensen et al. (2007) investigated the importance the top business schools attached to courses in the following three areas: ethics, corporate social responsibility (CSR), and sustainability. Their findings revealed four important aspects. First, a majority of the schools surveyed required that one or more of these topics be covered in their MBA curriculum and one-third of the schools require coverage of all three topics as part of their MBA curriculum. Second, there was a trend toward the inclusion of sustainability-related courses. Third, there was also a higher percentage of students interested in these topics. Fourth, several of the schools surveyed were teaching these topics using experiential learning and immersion techniques. Efforts such as those described above indicate the importance of ethics in management education and practice.

Recent advances in the study of business ethics are attempting to integrate a neural dimension into cognitions underlying ethical behavior and moral reasoning (Salvador and Folger, 2009). However, one must acknowledge that the neuroscience of ethics is different from the ethics of neuroscience. The former relates to the study of the neural foundations of ethical and moral behavior, whereas the latter deals with the ethical implications of using neuroscience methods to study the cognitive and neural mechanisms underlying ethical decision making. For Robertson,

Voegtlin, and Maak (2017) neuroscience can provide insight into individual reactions to ethical issues and raise challenging normative questions about the nature of moral responsibility, autonomy, intent, and free will.

Salvador and Folger (2009) discussed four themes emerging from neuroethics research. First, ethical decision making appears to be distinct from other types of decision making processes. Second, ethical decision making entails more than just conscious reasoning. Third, emotion plays a critical role in ethical decision making, at least under certain circumstances. Fourth, normative approaches to morality have distinct, underlying neural mechanisms. This is important because ethics is receiving increased attention in organizational scholarship and management practice. In fact, employees and managers encounter ethical issues and dilemmas on a regular basis.

1 UNDERSTANDING BUSINESS ETHICS AND MORALITY

1.1 Understanding Business Ethics

Ethics refers to the set of values and principles held by individuals and groups that influence their behaviors (Robertson et al., 2007). This set of values and principles leads people to engage in actions that involve moral issues that encompass ethical decisions. Scholars studying ethics contend that four stages are important in understanding whether a given action is ethical or not (Rest et al., 1999; Reynolds, 2008; Hannah, Avolio, and May, 2011). These stages are: (1) ethical awareness; (2) formulation of ethical judgments; (3) formulation of ethical intentions; and (4) acting ethically.

Ethical awareness involves whether an individual recognizes an issue as an ethical one. For example, a manager may contemplate hiring the protégé of his or her boss for an internship. The manager may not recognize this decision as an ethical one since it is just an internship. Becoming morally aware entails "identifying what we can in a particular situation, figuring out what the consequences to all parties would be for each line of action, and identifying and trying to understand our own gut feelings on the matter" (Rest, Bebeau, and Volker, 1986, p. 3). Rest also adds that "the person must have been able to make some sort of interpretation of the particular situation in terms of what actions were possible, who (including oneself) would be affected by each course of action, and how the interested parties would regard such effects on their welfare" (Rest et al., 1986, p. 7). For Butterfield, Trevino, and Weaver (2000), moral awareness refers to "a person's recognition that his or her potential decision or action could affect the interests, welfare, or expectations of the self or others in a fashion that may conflict with one or more ethical standards" (p. 982). Hence, the recognition of an issue as an ethical one is the first step in the ethical decision making process.

Once the person realizes that a situation is an ethical one, he or she may formulate an ethical judgment in deciding whether a given course of action is right or wrong. Hence, an ethical judgment is an evaluative statement regarding the ethicality of a given issue. Evaluative statements related to the wrongfulness or rightfulness of actions are rooted in moral judgments. This is particularly important because values and judgments play a critical role in making ethical decisions. Ethical judgments could lead to an overt intention for action or inaction. If the person recognizes the behavior as potentially ethical, he or she will act; otherwise, the person will refrain from acting. This assumption assumes that ethical intentions lead to actual ethical behavior. However, there may be situations where an individual may have genuine ethical intentions but may not follow through. This can happen in an organization where the prevailing culture does not always condone ethical practices and where employees are allowed to take shortcuts to accomplish organizational objectives.

The four-stage model of ethics could lead to the exploration of the following questions. What is an ethical issue? And what is an ethical dilemma? An ethical issue is a problem, situation, or opportunity that requires an individual, group, or organization to choose among several actions that must be evaluated as right or wrong, ethical or unethical, whereas an ethical dilemma is a situation that requires an individual, a group, or an organization to choose among several actions that have negative outcomes. For example, in the trolley car dilemma (below) involving whether to switch a button to turn right – an action that would result in the death of one person – or go straight – which will result in the death of five people – both scenarios lead to bad outcomes. Yet, a person must still decide what to do.

Today, business ethics has become both a field of study and a practice. As a field of study, business ethics addresses issues of morality and ethics in the workplace. The practice of business ethics concerns whether specific business practices are acceptable or not. It also includes the development of ethical codes as guidelines for employee behavior in organizations. It is worth mentioning that business ethics is often controversial, and there is no universally accepted approach for resolving ethical issues. If business ethics is a practice, then whose ethical standards should prevail when organizations operate in different nations and cultures? Should there be a "universal" code of ethics for organizations, or should ethics be context dependent?

People rely on ethical ideologies when making ethical judgments. Those who use a subjective approach to ethics contend that there are no universal guidelines for ethical behavior and ethics is context specific. The old saying, "While in Rome do as the Romans do" better illustrates this ethical ideology. Ethical relativism poses serious challenges to the application of ethical principles in organizations. If ethics is subject to individual interpretation, then how can organizations arrive at a consensus to develop and implement ethical standards that can serve as moral code for their members? In addition, what ethical standards should prevail when organizations operate in different societies and cultures? A second ideology, idealism, views ethics as representing a set of universal principles that people can adopt and follow regardless of their background. Idealists argue that there are moral standards that people of all cultures hold and agree on. For example, all cultures agree that the unlawful murder of another human being is wrong and therefore should be avoided. For universalists, the best possible outcome can always be achieved by following universal moral rules (Forsyth, 1980, 1985).

Along with these two ethical ideologies, research on business ethics has also integrated utilitarian, deontological, and virtue ethics principles. Discussing each of these three theories in depth is beyond the scope of this book but, briefly, utilitarianism focuses on actions that benefit the greatest number of individuals, whereas the deontological approach calls on the sense of obligation and duty. The virtue ethics approach emphasizes character and personal "goodness" as essential in making ethical decisions. Virtue ethics is relevant to business ethics and argues that "ethical behavior involves not only adhering to conventional moral standards but also considering what a mature person with a 'good' moral character would deem appropriate in a given situation" (Ferrell, Fraedrich, and Ferrell, 2017).

Applying these three ethical theories in organizations would lead to different types of behaviors and probably different outcomes. A utilitarian approach would advocate that managers engage in actions that yield the greatest benefit for the greatest number of stakeholders. However, a deontological perspective would favor engaging in actions out of duty. Virtue ethics would imply that managers and employees cultivate virtues. By becoming virtuous people, they would transform organizations into moral communities. After reviewing the literature on the three ethical theories of utilitarianism, deontological, and virtue ethics, Slingerland (2011) concludes that the virtue ethics model of self-cultivation more accurately represents how real human beings engage in moral reasoning and is also better adapted as an educational technique to the evolved cognitive architecture of human beings than deontological or utilitarian approaches (p. 82).

1.2 Morality and Moral Judgments

Human morality provides the foundation for many of the pillars of society, informing political legislation and guiding legal decisions while also governing everyday social interactions (Funk and Gazzaniga, 2009). As Suhler and Churchland (2011, p. 2103) put it, "morality permeates human existence, playing a role in a vast array of choices and evaluations, both public and private, momentous and trifling." Moral judgments depend on information about agents' beliefs and intentions (Young and Dungan, 2012). Morality refers to the "interlocking sets of values, practices, institutions, and evolved psychological mechanisms that work together to suppress or regulate selfishness and make social life possible" (Haidt, 2008, p. 70).

"Philosophers and social scientists have long sought to understand the nature of human moral systems" (Hitlin and Vaisey, 2013, p. 51). Hitlin and Vaisey (2013) contend that "in social science, the word moral is used in two distinct senses. First, moral refers to correspondence with universal standards of right and wrong linked to concerns about justice, fairness, and harm. Second, moral refers to understandings of good and bad, right and wrong, worthy and unworthy that vary between persons and between social groups" (p. 55).

Definitions of "moral judgments" include "judgments of the rightness or wrongness of acts that knowingly cause harm to people other than the agent" (Borg, Hynes, Van Horn, 2006, p. 803). Moral judgments are broadly defined as evaluative judgments of the appropriateness of one's behavior within the context of socialized perceptions of right and wrong (Moll et al., 2005). Tomasello and Vaish (2013) suggest that from an evolutionary perspective, morality is a form of cooperation that requires individuals either to suppress their own self-interest or to equate it with that of others. Greene (2015) also notes that morality promotes cooperation.

Human moral judgments depend on "theory of mind," which is the capacity to represent the mental states of others (Adolphs, 2003). Theory of mind "underlies our ability to empathize with others, to judge how they might react in response to our actions, and to predict the subjective consequences of our actions for conspecifics" (Casebeer, 2003, p. 844). People attribute mental states to agents performing certain tasks. When people judge the morality of the actions of others, they tend to make "moral inferences." For example, when people believe that the agent acted genuinely without malice, they may tend to exonerate them. However, when they believe that an actor acted knowingly to harm others, they will tend to consider the action as immoral and perhaps punish them. Understanding morality is important for organizations. Indeed, organizations engage in activities that do have consequences for individuals and communities. Hence, acting morally could help organizations reduce the potential negative impact of their activities on various stakeholders.

1.2.1 Moral dilemmas

Moll, De Oliveira-Souza, and Eslinger (2003, p. 300) define a moral dilemma as "a problem situation in which dissonant moral emotions of roughly comparable strength are elicited, giving rise to a slow and effortful process often referred to as moral reasoning." "Moral behavior stems from a delicate balance between prosocial and altruistic behaviors at one extreme, and antisocial and selfish behaviors, at the other" (ibid.). Greene et al. (2001) argue that moral dilemmas vary systematically in the extent to which they engage emotional processing and that these variations in emotional engagement influence moral judgment. They use the trolley car dilemma as an example. In the trolley car dilemma, a driver is headed toward five road workers who will be killed unless the driver pushes a switch to direct the trolley car toward a track with only one road worker. Going straight would result in the death of five people, whereas turning right will only lead to the death of one person. Asked the question of the best course of action to take, most people choose to turn right, thereby choosing to kill one person. The rationale behind the switch decision is that it is better to kill one person than to kill five people.

Another version of the trolley car dilemma is the footbridge dilemma. In this scenario, the person is standing on a footbridge as an onlooker. The trolley car is racing toward five people who would surely die if something is not done. As an onlooker, you suddenly notice a very big man standing on the bridge. If the big man is pushed, he will fall on the track on front of the trolley car and prevent the death of the five people but will die. Faced with this dilemma, most people prefer not to shove the big man. Greene and his colleagues argue that the two situations are quite different and appeal to two different schools of philosophical thought (Greene et al., 2001). The first scenario results from deliberate reasoning and entails the use of utilitarianism or consequentialism principles. The second scenario, however, is emotionally driven and is best explained by the deontological perspective. According to Greene

(2009), "characteristically deontological judgments are driven by automatic emotional responses, while characteristically utilitarian judgments are driven by controlled cognitive processes" (p. 581).

How people judge the morality of actions can also depend on whether these actions are perceived as deliberate or not. Deliberate actions that harm others or deprive them of some valuable benefits may tend to be viewed as immoral compared to those that might be considered as part of unintended consequences or side-effects. Cushman (2013) notes that "harmful actions are judged more severely when used as a means to accomplish a goal than when brought about as a side-effect of accomplishing a goal" (p. 275).

1.2.2   The dual system framework for morality

Early theories of morality and moral judgments emphasized a rational perspective in which moral reasoning played an important role in moral behavior (Kohlberg, 1969). The rational approach in moral judgments assumes that moral behavior is reached through reasoning and reflection (ibid.). An individual will make a moral decision after consciously deliberating about the facts involved in the situation. However, there is mounting evidence that moral judgments are emotion driven (Haidt, 2008, 2012; Greene, 2008). Haidt (2001) developed a social intuitionist model, which contends that moral reasoning is rarely the direct cause of moral judgment. "The central claim of the social intuitionist model is that moral judgment is caused by quick moral intuitions and is followed (when needed) by slow, ex post facto moral reasoning" (p. 817).

Research in moral psychology and other social sciences has underscored the existence of a dual process (Lieberman et al., 2002; Kahneman, 2003; Satpute and Lieberman, 2006). On one hand, there is an emotional, automatic process, and on the other, a controlled, systematic process. Hence, morality can be viewed as combining both emotional and controlled processes. Consequently, affective and cognitive components of morality must be integrated (Cushman, 2013). Hence, moral judgment is influenced by both automatic emotional responses and controlled, conscious reasoning (Greene, 2009; 2014). Greene (2014) notes that the human brain faces a tradeoff between efficiency and flexibility, and to promote efficiency the brain has point-and-shoot automatic settings in the form of intuitive emotional responses. This perspective echoes Haidt's (2001, 2007) social intuitionist model, which contends that moral reasoning is rarely the direct cause of moral judgment; rather, it is caused by quick moral intuitions followed by slow, ex post facto moral reasoning (Haidt, 2001, p. 817).

According to Grasso (2013), the dual-process theory indicates that up-close, and personal harm triggers deontological moral reasoning, whereas harm originating from impersonal moral violations, like those produced by climate impacts, prompts consequentialist moral reasoning. Orlitzky (2017) suggests that a dual-process explanation of emotional-intuitive automaticity (reflexion) and deliberative reasoning (reflection) is the most appropriate view. However, the author contends that the evidence contradicts Greene's conclusion that non-consequentialist decision making is primarily sentimentalist or affective at its core, while utilitarianism is largely rational-deliberative. Hence, "a growing number of cognitive scientists and philosophers have come to agree with David Hume and the Stoics that normative judgments are ultimately derived from human emotional reactions" (Slingerland, 2011, p. 89).

For Van Bavel, Hall, and Mende-Siedlecki (2015), "moral cognition emerges from the integration of and coordination of a widely distributed set of brain regions" (p. 170). Consequently, morality is a dynamic system and therefore cannot be confined to a dual process of intuition versus reason. This echoes Gino's (2015) suggestion that morality is dynamic and is not a stable trait that characterizes individuals. Moreover, morality is not a unified psychological or neurological phenomenon (Borg, Lieberman, and Kiehl, 2008). Despite this criticism, there are some benefits in exploring the neural basis of ethics and morality.

2 NEURAL BASIS OF BUSINESS ETHICS AND MORALITY

Scholars use the term neuroethics to discuss the neural basis of ethical decisions as well as the ethical implications of using neuroscience to explain human behavior. In fact, "the word 'neuroethics' entered the vocabulary of academic neuroscientists and bioethicists at the beginning of the twenty-first century" (Farah, 2012, p. 572). According to Farah (2012), US political columnist William Lewis Safire first coined the term in the *New York Times* and it has come to refer to a broad range of ethical, legal, and social issues raised by progress in neuroscience. It also refers to the study of the neural basis of judgments of right and wrong and ethical behavior. Hence, I use two terms – the ethics of neuroscience and the neuroscience of ethics – to explain these two perspectives.

2.1   The Ethics of Neuroscience

As indicated earlier, neuroethics can have two meanings. First, neuro-ethics can be viewed as a means to understand how neuroscience could help explain how we make ethical and moral decisions. Second, neuro-ethics can be described as raising questions about the ethicality of applying neuroscientific methods to understand human behavior. In this regard, the term neuroethics emerged as a way to draw attention to ethical issues concerning different aspects of brain research (Olteanu, 2015). For example, is it ethical to "invade" the brain to identify brain regions that could help explain behaviors such as leadership and decision making? Is it ethical to attempt neurological interventions to correct what could be described as "neurological deficiencies," when in fact partici-pants do not show any sign of illness?

These questions pose an ethical dilemma related to the use of neuroscientific research to improve employee behavior in the workplace. In this second conceptualization, neuroethics is defined as "being con-cerned with ethical, legal, and social policy implications of neuroscience, and with aspects of neuroscience research itself" (Illes and Bird, 2006, p. 511). This definition is similar to the one put forth by Farah (2012), who defines neuroethics as consisting of "a broad range of ethical, legal, and social issues raised by progress in neuroscience" (Farah, 2012, p. 572). Studying the ethical implications of neuroscience is not limited to the organizational sciences; it has also been a concern for marketing scholars interested in the use of neuroscientific tools to understand customer behavior (Olteanu, 2015). Frost and Lumia (2012) contend that the issue is not simply the ethics of neurotechnology, but also the incorporation of neuroscientific findings into a richer understanding of human ethical functioning. The ethics of neuroscience could help researchers avoid the pitfalls that may be related to conducting research on healthy brains that may influence people's behavior in organizations.

A case in point concerning neuroethics is the use of brain-enhancement techniques. "Brain enhancement refers to interventions that make normal, healthy brains better, in contrast with treatments for unhealthy or dysfunctional brains" (Farah, 2012, p. 579). For example, "coffee, tea, coca leaves, and alcohol are among the familiar substances used to alter brain chemistry for improved cognition or mood" (ibid.). To what extent can the use of these techniques to enhance brain power be considered unethical? Relying on the most common of these substances, such as coffee and tea, to augment brain power could be considered as less controversial. However, relying on brain modification techniques that augment the brain's capacity to process information can raise ethical questions if it leads to some people performing better than others. It could constitute some form of cognitive doping analogous to athletes using performing-enhancing drugs.

Lindebaum (2013a) argues that using neuroscientific methods and principles to identify and develop high-performing leaders is an ethically sensitive issue. For example, neuroscientific studies of leadership may indicate that some managers lack leadership skills as a result of some "brain limitations." Does this lead to some type of intervention? An ensuing concern is whether neuroscientific tools should be used to train managers to perform effectively in the absence of neural deficiencies. Lindebaum (2013b) addresses the ethical implications of neuroscientific methods to study management behavior. In particular, he identifies the concept of "brain profile deficiencies" coined by Waldman, Balthazard, and Peterson (2011a, 2011b) as a major limitation in using neuroscience to study organizational behavior and provide guidelines for management practice. Lindebaum (2013b) believes that neuroscientific studies applied in organizations face ethical issues not only because of the tools they use but because of the conclusions they reach and the recommendations they make. To remedy this limitation, he suggests that organizational neuroscience should be construed as one of the disciplines of positive organizational scholarship.

According to Roskies (2002), the ethics of neuroscience can be roughly subdivided into two groups of issues: (1) the ethical issues and considerations that should be raised in the course of designing and executing neuroscientific research; and (2) the evaluation of the ethical and social impact that the results of those studies might have. Roskies (2002) calls the first part "the ethics of practice" and the second the "ethical implications of neuroscience" (p. 21). One of the tools used by neuroscientific research is functional magnetic resonance imaging (fMRI), which presents ethical challenges (Farah, 2005).

Although organizational neuroscience can lead to new questions on the neural basis of ethics and morality in organizations, the discipline itself faces ethical issues of its own (Lindebaum, 2013b; Cropanzano and Becker, 2013). Cropanzano and Becker (2013) argue that "neuroscience is a potentially powerful scientific tool of organizational research, but in addition to its great promise, this emerging discipline also confronts us with ethical risks" (p. 306). Hence, it is important to make a distinction between the ethics of neuroscience and the neuroscience of ethics. Some researchers study the ethics of conducting neuroscientific research, whereas others study "the neuroscience of ethics," referring to the exploration of how brains make decisions when confronted with moral dilemmas. Both approaches are equally valid for organizational research (Greely, 2007).

2.2 The Neuroscience of Ethics

The term neuroscience of ethics is used here to refer to the study of the neural foundations of ethical behavior. Reynolds (2006) proposed a neurocognitive model of ethics in which he emphasized the role of neurons in explaining ethical decisions. He notes that "cognitive models with their insistence on linear progression or movements from stage to stage, are incapable of explaining why managers 'know,' without being able to explain why, that a particular course of action is ethical and another is not" (Reynolds, 2006, p. 742). His model suggests that ethical decision making involves two interrelated yet functionally distinct cycles – a reflexive pattern-matching cycle and a higher-order conscious reasoning cycle – and thereby describes not only reasoned analysis, but also the intuitive and retrospective aspects of ethical decision making. His model is consistent with the dual-process model of moral judgment described earlier.

The neuroscience of ethics may help organizational scholars understand how employees and managers make decisions that bear moral components. Such understanding could provide insights and guidelines that could help avoid ethical blunders in organizations. For Farah (2012, p. 573), "any endeavor that depends on being able to understand, assess, predict, control, or improve human behavior, is, in principle, a potential application area of neuroscience. This includes diverse sectors of society, for example, education, business, politics, law, entertainment, and warfare." Hence business ethics is an area where research in neuroscience can advance knowledge and inform practice.

Neuroscientific evidence indicates that "moral judgment is often an intuitive, emotional matter" (Greene, 2003, p. 849). Moll et al. (2002) demonstrate that viewing moral and non-moral pleasant visual stimuli activates a common network of brain areas that includes the amygdala, the orbitofrontal cortex (OFC), the medial frontal gyrus (MedFG), and the cortex surrounding the right posterior superior temporal sulcus (pSTS). Thus, people may develop "moral prototypes" against which incoming stimuli are compared. Moll, de Oliveira, and Zahn (2008) found that the dorsolateral prefrontal cortex (dlPFC), the anterior prefrontal cortex (aPFC), the medial prefrontal cortex (mPFC), the ventromedial prefrontal cortex (vmPFC) and the superior temporal sulcus (STS) region are implicated in moral cognition and behavior. Ryan (2017) notes that results from neuroscience studies, especially those that focus on the differential effects of oxytocin and testosterone, on males and females, could benefit six areas of business ethics research: trust, moral decision-making, organizational justice, moral development, the ethics of care, and female management styles. Greene (2003) argues that neuroscience can have profound ethical implications by providing us with information that will prompt us to re-evaluate our moral values and our conceptions of morality.

2.3　The Neural Basis of Moral Judgments

De Waal (1996) contends that morality is as firmly grounded in neuro-biology as anything else we do or are. He argues that "once thought of as purely spiritual matters, honesty, guilt, and the weighing of ethical dilemmas are traceable to specific areas of the brain" (De Waal, 1996, pp. 217–218). Young and Dungan (2012) note that morality is virtually everywhere in the brain. Efforts to explore the neural basis of morality and moral judgments have contributed to the emergence of the field of moral cognitive neuroscience, which is an interdisciplinary field based on the integration of psychology, neuroscience, evolutionary biology, and anthropology and that "aims to elucidate the cognitive and neural mechanisms that underlie moral behavior" (Moll et al., 2005, p. 799).

For Smetana and Killen (2008), concerns for the neural foundations of morality have given rise to the emergence of moral neuroscience, which seeks to establish a biological basis for moral judgments and emotions. Although scholars cannot identify a moral center of the brain, they can still describe some brain structures that are activated when people deal with issues of ethics and moral judgments. According to Moll et al. (2003), from a neuroscience perspective, a cortico-limbic network is recruited during the performance of moral judgments. This network includes the orbitofrontal, prefrontal, the anterior temporal and the anterior cingulate cortexes, the thalamus, and the midbrain regions. However, the neural systems implicated in morality are also involved in other cognitive processes (Decety and Cowell, 2014).

Greene et al. (2004) found that difficult as compared to easy personal moral dilemmas involved increased activity bilaterally in both the dlPFC and the inferior parietal lobes. Moral dilemmas were also associated with increased anterior cingulate cortex and posterior cingulate cortex activity. According to Greene et al. (2004), moral judgments can be personal or impersonal. Personal moral judgments are driven by emotion, whereas impersonal moral judgments are driven by cognition. Greene et al. (2004) note that a moral violation is personal if it meets three criteria: (1) the violation must be likely to cause serious bodily harm; (2) the harm must befall a particular person or set of persons; and (3) the harm must not result from the deflection of an existing threat to a different party.

Young and Saxe (2008) showed that brain structures such as the medial prefrontal cortex, play an important role in moral judgment. Using transcranial magnetic stimulation (TMS), Sacco et al. (2017) found that participants who received anodal stimulation assigned less blame to accidental harms compared to those who received cathodal or sham stimulation. Borg et al. (2011) found hemodynamic activity in the bilateral anterior insula and basal ganglia that correlates with the moral verdict, "this is morally wrong," as opposed to, "this is morally not-wrong." Using comparisons of deliberation-locked versus verdict-locked analyses, the authors also demonstrated that hemodynamic activity in high-level cortical regions previously implicated in morality, including the ventromedial prefrontal cortex, the posterior cingulate cortex, and the temporoparietal junction, correlated primarily with moral deliberation as opposed to moral verdicts.

A review of the literature on the neural basis of morality led Forbes and Grafman (2010) to conclude that the prefrontal cortex plays a significant role in moral judgment and social cognition. Cushman et al. (2012) explore the tendency to judge harmful actions morally worse than harmful omissions (the "omission effect") using fMRI. Because ordinary people readily and spontaneously articulate this moral distinction, it has been suggested that principled reasoning may drive subsequent judgments. If so, people who exhibit the largest omission effect should exhibit the greatest activation in regions associated with controlled cognition. Cohen (2005) notes that personal moral dilemmas activate the medial prefrontal cortex associated with emotional arousal, whereas impersonal moral dilemmas elicit activity in the prefrontal cortex, which is associated with cognitive processes. Prehn et al. (2008) contend that both cognitive and emotional components play an important role in formulating moral judgments.

Moral sensitivity is the ability to detect and evaluate moral issues. By detecting a moral issue, a person can determine whether a moral action should be taken. Moral sensitivity is construed as the prerequisite for moral actions (Rest, 1984). Using fMRI and contextually standardized, real-life moral issues, Robertson et al. (2007) demonstrated that sensitivity to moral issues was associated with activation of the polar medial prefrontal cortex, dorsal posterior cingulate cortex, and posterior superior temporal sulcus. These findings indicate the neural underpinnings of moral sensitivity. However, more research is needed, specifically in assessing individual differences. For example, researchers may compare individuals who are morally more sensitive to those who are less so and determine whether some brain regions are more activated for the most morally sensitive participants compared to those who are morally less sensitive.

Moretto et al. (2010) contend that research evidence suggests that emotion processing mediated by the ventromedial prefrontal cortex (vmPFC) is necessary to prevent personal moral violations. They found that when facing moral dilemmas, patients with lesions in the vmPFC are more willing than normal controls to approve harmful actions that maximize good consequences (e.g., utilitarian moral judgments), yet, none of the existing studies had measured participants' emotional responses while they considered moral dilemmas. In their study, Moretto et al. (2010) had patients suffering from lesions in the vmPFC and control participants consider moral dilemmas while skin conductance response (SCR) was measured as a somatic index of affective state. They found that vmPFC patients approved more personal moral violations than did controls. The authors also found that, unlike control participants, vmPFC patients failed to generate SCRs before endorsing personal moral violations. In addition, such anticipatory SCRs correlated negatively with the frequency of utilitarian judgments in normal participants. These findings provide direct support for the hypothesis that the vmPFC promotes moral behavior by mediating the anticipation of the emotional consequences of personal moral violations.

Nash, Baumgartner, and Knoch (2017) contend that group-focused moral foundations are moral values that help protect the group's welfare. They found that increased adherence to group-focused moral foundations was strongly associated with reduced gray matter volume in key regions of the conflict detection and resolution system (anterior cingulate cortex and lateral prefrontal cortex). Cameron et al. (2017) contend that implicit moral evaluations that are immediate, unintentional assessments of the wrongness of actions or persons play a central role in supporting moral behavior in everyday life. Luo et al. (2006) found that implicit moral attitude was associated with increased activation in the right amygdala and the ventromedial orbitofrontal cortex. As indicated previously, an understanding of the neural basis of ethics and morality could have implications for ethical behavior in organizations.

3   IMPLICATIONS FOR ORGANIZATIONS

Organizational members face ethical issues and moral dilemmas on a regular basis. Moral dilemmas such as the one facing a driver of a trolley car and a bystander on a footbridge could also be expanded to business settings. For example, to meet an order a manager may decide not to inspect all products before shipping them to a buyer. Another manager may hide the defective parts of a product or even misrepresent the true value of a company to prevent a slide in its stock price. All these situations raise ethical issues for managers and employees.

The work environment imposes physical, social, and moral demands on employees. Hence, organizations are entities that put pressure on members to adopt certain desirable behaviors, thereby putting some "morality weight" on them. For these reasons, organizations can be construed as moral communities (Bowie, 1999; Haidt, 2007). For example, drawing on the philosophy of Immanuel Kant, Bowie (1999) argues that business organizations are moral communities. Like Immanuel Kant, Bowie's perspective focuses on deontology, the sense of duty and obligation. Today, organizations codify their moral compass in the form of ethical codes. The existence of an ethical code can lead to the emergence of three moral components: moral intensity (Jones, 1991; Barnett, 2001), moral approbation (Jones and Ryan, 1997, 1998), and moral outrage (Bies and Moag, 1986; Montada and Schneider, 1989). Jones (1991) defines moral intensity as the extent of issue-related moral imperative in a situation (p. 372). In an organization, managers may consider some issues as conveying more importance than others. For example, an organization that subcontracts manufacturing in Southeast Asia may consider the use of child labor and sweatshops as a moral issue that must be at the top of its agenda. Such a focus could guide managers' actions.

Jones and Ryan (1997) defined moral approbation as moral approval from oneself or others. They contend that people consider four factors when determining their own or someone else's level of moral responsibility: (1) the severity of the act's consequences; (2) the certainty that the act is moral or immoral; (3) the actor's degree of complicity in the act; and (4) the extent of pressure the actor feels to behave unethically (p. 663). Whether an individual engages in a given action would depend on the degree to which the action meets the moral threshold. To some extent, moral approbation can be construed as a form of social validation, especially when people care about whether their actions are morally approved (or not) by others. In fact, people rely on the opinions of their referent groups when deciding how to behave. People evaluate their actions against moral thresholds held by their referent groups. The existence of a code of ethics may also lead employees to expect others to see them as following the ethical guidelines prescribed by the organization. Consequently, they may want to be seen as ethical and moral agents.

Hannah and Waldman (2015) contend that neuroscience can contribute to moral approbation by allowing group members to "express other moral emotions, such as disgust or anger toward a member who violates norms, which could result in the experience of self-focused emotions, such as shame, guilt, or fear of condemnation, experienced by the transgressor" (p. 245). Mirror neurons may probably explain the spread of ethical or unethical behavior in the workplace. Perhaps, witnessing others act ethically may unconsciously compel others to do the same. Conversely, witnessing others act unethically may lead others to act likewise.

Organizational members may experience moral outrage when they witness ethical violations. Moral outrage refers to the anger provoked by the perception that a moral standard has been violated (Bies and Moag, 1986; Montada and Schneider, 1989). It is often the precursor to moral motivation (Haidt, 2003). Moral motivation may compel employees who witness wrongdoing to take action to punish the perpetrators of the perceived unethical conduct. Whistleblowing may be an illustrative example of moral motivation driving others to go public about wrong-doings by their employers. Moral sensitivity may perhaps lead some individuals to be more outraged by their company's violations of ethical standards than others.

Despite efforts to codify ethical processes in organizations, organizational processes such as hierarchy, bureaucracy, structure, culture, and the pressure to perform may impact members' moral compass. "Cheating, deception, organizational misconduct, and many other forms of unethical behavior are among the greatest challenges in today's society" (Gino, 2015, p. 107). In fact, unethical workplace behaviors can have far-reaching consequences, including job losses, risks to life and health, psychological damage to individuals and groups, social injustice and exploitation and even environmental devastation (Lindebaum, Geddes, and Gabriel, 2016).

In organizations there is often a gap between the desire to have a positive moral self-image and dishonest behavior. Hence, organizational members may often find a disconnect between their personal values and those of their organization. The circumstances surrounding certain decisions may also influence the desire for congruence between ethical values and actual behavior. According to Chakroff et al. (2016), "a reasonable predictor of how people may behave in the future is how they behaved in the past" (p. 201). Hence, organizations may be well served by considering the ethical background of the employees and managers they hire. Doing so could contribute to the development of an ethics culture.

Organizations could also train their members to improve their moral capacity. Doing so could help improve what Hannah et al. (2011) called moral maturation and moral conation. Moral maturation refers to the "capacity to elaborate and effectively attend to, store, retrieve, process, and make meaning of morally relevant information" (p. 667). It includes moral identity, complexity, and metacognitive ability. Moral conation is the "capacity to generate responsibility and motivation to take moral action in the face of adversity and persevere through challenges" (p. 664). It includes moral courage, efficacy, and ownership. According to Hannah et al. (2011), both moral maturation and moral conation can help ethical decision making.

How can neuroscience help organizations solidify their status as moral communities? Can managers and employees be trained using neuroscience techniques to become more ethical? In addressing this question, Hannah and Waldman (2015) suggest the use of neurofeedback as an appropriate neuroscientific tool. They suggest that neurofeedback could help to improve employees' capacity to make moral decisions. Specifically, it can enhance their capacity for moral maturation and moral conation. However, they also point out that enhancing moral capacities or motivations cannot always translate into acting morally in face of challenging situations.

How do people decide what is right and wrong? What brain structures are involved in this process? Perhaps, when people are familiar with how their own brains work, they may pay attention to some of their decisions. Although some brain structures are implicated in ethical and moral decision making, it is still premature to pinpoint an ethical and moral area of the brain. This should not, however, prevent organizational neuroscience scholars from exploring neuroscience techniques that could prove useful in helping employees become more ethical.

10. The neural basis of emotions and unconscious bias

Organizations are arenas of affect production, that is, environments in which people interact with one another, work with or for others, and experience a variety of emotional states. Reina, Peterson, and Waldman (2015, p. 227) note that "organizations are comprised of individuals who are filled with emotions that are influenced or expressed on a regular basis in reaction to people, events, and circumstances around them." Such interactions produce opportunities for emotional expression and contagion. Hatfield et al. (1994) define emotional contagion as the automatic and unconscious transfer of emotions. Indeed, people are affected by the emotions of others without conscious effort. Hence, emotional transfer can be a subtle way of influencing others.

Psychologists and organizational behavior scholars have studied the types and sources of human emotions, as well as their impact on actual behavior (Hatfield et al., 1994; Grandey, 2000). Such studies were motivated by the fact that understanding human emotions could help regulate them. This goal is particularly important because "the ability to regulate our emotional responses and states is a critical component of formal social function and adaptive interactions with the environment" (Phelps, 2006, p. 44).

Classical economics emphasizes the rational choice model. Descartes's "*cogito ergo sum*" (I think, therefore I am) has led to the neglect of the role of emotion in explaining human decision making. However, research in psychology and recently, behavioral economics and neuroeconomics, has presented evidence to the contrary by demonstrating that there is an interaction between emotion and cognition (Phelps, 2006; Lerner et al., 2015). Lerner et al. (2015) argue that emotions and decision making go hand in hand and that "emotions powerfully, predictably, and pervasively influence decision making" (p. 802).

Thus, understanding emotions in organizations is important because social interactions are prone to emotional reactions. For example, we like or dislike others based on their behavior and characteristics, but also based on our own mental models. People experience anxiety and fear in the workplace on a regular basis. According to LeDoux (2000), emotional arousal has a powerful influence over cognitive processing, and attention, perception, memory, decision making and the conscious concomitants of each are all swayed by emotional states because emotional arousal organizes and coordinates brain activity. We know that these emotions are rooted in biology and neuroscience. Hence, it is important to understand the neural basis of emotions and emotional regulation because "knowledge of the role of emotions in changing a given brain state will no doubt help in improving economic predictions" (Martins, 2010, p. 5).

1 NEURAL FOUNDATIONS OF EMOTIONS

The study of the neural basis of emotion has led to the development of the field of affective neuroscience, which is the study of the neural basis of affect in humans and non-humans (Davidson, Jackson, and Kalin, 2000; Ochsner and Gross, 2005). Indeed, "emotions modulate and guide behavior as a collection of biological, social, and cognitive components" (Deak, 2011, p. 71). Recent research has explored the neural basis of emotions (Phan et al., 2004; Vul et al., 2009). Such research highlights the effects of some brain structures on emotions, the neural basis of emotional regulation, and whether specific emotions such as empathy have neural underpinnings. "The experience of emotion is a system-level property of the brain" (Barrett et al., 2007, p. 3).

Understanding the neural basis of emotions has implications for organizations. It is commonly accepted that emotions change our thinking patterns. For example, when people are calm, the frontal lobes guide slow and rational thinking, described as cold cognitions (Phan et al., 2004; Vul et al., 2009). However, when people are angry or stressed, hot cognitions prevail. Understanding this difference could help employees and managers determine when to angrily respond to an offending e-mail or cool down before responding. Doing so could help one keep one's amygdala in check. I have personally experienced situations where colleagues have sent hot-headed e-mails to others including peers, subordinates, and superiors, possibly signaling impulsivity, or a lack of maturity or self-control.

Very often, employees experience emotional dissonance in the workplace. Emotional dissonance is the conflict originating from expressed and experienced emotions (Abraham, 1998). Despite this emotional dissonance, employees are expected to display the required emotions (often impossible emotions), thereby leading to emotional labor (Hochschild, 1983; Grandey and Gabriel, 2015). Emotional labor refers to the "management of feeling to create a publicly observable facial and bodily display" (Hochschild, 1983, p. 7). "The management of felt and displayed emotions is an important aspect of many employees' jobs, particularly in service industries where the expression of positive emotions is an expected part of service delivery" (Pugh, Groth, and Hennig-Thurau, 2011, p. 377). Neuroscience could help us to better understand emotional labor in organizations (Reina et al., 2015).

Emotional labor includes two components: surface acting and deep acting. Surface acting consists of modifying one's behavior to be consistent with expected emotions while continuing to hold different internal feelings. For example, an employee may be angry at a disgruntled customer, but still smile and address his or her request. Deep acting consists of modifying one's behavior to match the required emotions. A manager can try to brush off irritation caused by a disgruntled employee and provide him or her with some advice. Reina et al. (2105) suggest that neuroscience could help to better understand both surface acting and deep acting, and consider reinterpreting and distancing as two ways to manage surface and deep acting. In fact, using the two types of emotional labor involves some form of emotional regulation. To the extent that people are able to "reappraise" an emotionally charged situation, they could turn negative emotions into positive or at least neutral ones. Likewise, when people distance themselves from a situation, they tend to be less negatively affected. Distancing can also help to better cope with emotionally charged situations.

1.1 Neural Structures Involved in Emotions

Emotions play an important role in decision making and other types of reactions (Damasio, 1994; Bechara and Damasio, 2005; Glimcher, Kable, and Louie, 2007). For example, Pillutla and Murnighan (1996) and Van Winden (2007) suggest that punishment in the Ultimatum Game is more driven by anger about the appropriation of resources than a concern for fairness, indicating that emotions guide such behavior. Phelps (2006) notes that the neural circuitry of emotion and cognition interact from early perception to decision making and reasoning. The amygdala is one brain structure that is often implicated in emotional reactions. The amygdala is a brain structure implicated in many kinds of phenomena such as attitudes, stereotyping, person perception, and emotion (Ochsner and Lieberman, 2001) and can be construed as the seat of human emotions. Brain structures such as the amygdala, the anterior cingulate cortex and the prefrontal cortex are all implicated in emotions.

1.2 Neural Basis of Emotional Regulation

The capacity to control emotion is important for human adaptation (Ochsner and Gross, 2005, p. 242). Emotion regulation can be understood as a hierarchical control system that, at various levels, modulates autonomic reactions, appraisal mechanisms, attention, the contents of working memory, and goal-directed action selection (Smith and Lane, 2015). Perceiving one's own emotions involves a multi-stage interoceptive/somatosensory process by which these body state patterns are detected and assigned conceptual emotional meaning (ibid.). According to Ochsner and Gross (2005) there are two forms of emotional regulation: (1) controlling attention to stimuli; and (2) cognitively changing the meaning of emotionally evocative stimuli (re-appraisal). Ochsner and Gross (2005) showed that these two forms of emotion regulation depend on the interactions between prefrontal and cingulate control systems and cortical and subcortical emotion-generative systems. Using either form could help reduce emotional intensity. For example, research on emotional regulation has demonstrated that expressing one's feelings reduces the "pain" of these feelings by diminishing activity in the amygdala (Ochsner and Gross, 2005; Gainotti, 2012). This reduction of activity diminishes the negative impact of these emotions.

Lieberman et al. (2007) found that affect labeling, or putting feelings into words, diminished the response of the amygdala to negative emotional images. They also found that affect labeling produces increased activity in the right ventrolateral prefrontal cortex (rvlPFC) and an inverse correlation between activity in the amygdala and the rvlPFC. This could be important in helping employees cope with negative stimuli and frustration at work. Putting feelings into words is one of the best ways of managing negative emotional responses, thus employees could be given the opportunity to vent their frustrations without fear of retribution. Opportunities to voice personal feelings and frustrations could lead to a form of "psychological release."

Using functional magnetic resonance imaging (fMRI), Phan et al. (2004) found that the nucleus accumbens responded to both increasing emotional intensity and self-relatedness. They also found that activity in the amygdala was specifically related to affective judgments and emotional intensity and that appraisal activated the ventral medial prefrontal cortex, the dorsal medial frontal cortex, and the insula. Emotions such as fear could be construed as adapted responses to environmental stimuli. In fact, accurate appraisal of environmental signals is central to an organism's survival (Phan et al., 2004, p. 768). LeDoux (2003) focuses on the neural bases of fear conditioning and concludes that the amygdala plays an important role.

Phelps, Lempert, and Sokol-Hessner (2014) note that changing emotions can alter choices and there is no clean delineation between brain regions underlying emotion and cognition. They also argue that "there is no clear evidence for a unified system that drives emotion" (Phelps et al., 2014, p. 265) and suggest that "one could argue that choice itself is indicative of an affective response because it signals an evaluation of preference, motivation, or subjective value assigned to the choice options" (p. 267). The authors argue that emotion influences choice through incidental affect. They define incidental affect as a baseline affective state that is unrelated to the decision itself (ibid.). Examples of incidental affect include stress and mood. "In our everyday lives, the stimuli most likely to elicit emotional responses are other people" (Phelps et al., 2014, p. 275). This is particularly true for organizations where employees interact on a regular basis.

Buhle et al. (2014) note that "cognitive reappraisal is an emotional strategy that involves changing the way one thinks about a stimulus in order to change its affective impact" (p. 2981). They conducted a meta-analysis of 48 studies mostly involving downward regulation of negative affect. Cognitive reappraisal consists of changing one's interpretations of affective stimuli (Buhle et al., 2014). "The implementation of reappraisal consistently activated cognitive control regions, including DMPFC, DLPFC, VLPFC, and posterior parietal lobe" (Buhle et al., 2014, p. 2984). Buhle et al. (2014) found that reappraisal consistently (1) activated cognitive control regions and lateral temporal cortex but not the ventromedial prefrontal cortex; and (2) modulated the bilateral amygdala, but not other brain regions. Cognitive reappraisal of emotions involves the use of cognitive control to modulate semantic representations of an emotional stimulus.

Harle et al. (2012) analyzed the effect of incidental affect on decision-making and found that people who have been induced to feel negative emotions such as sadness tend to reject more unfair offers in the Ultimatum Game. Receiving unfair offers while in a sad mood activated the anterior insula and the anterior cingulate cortex. In addition, sad participants showed a diminished sensitivity in the ventral striatum, a region associated with reward processing. Mental representations of emotions are associated with brain structures including the amygdala, the orbitofrontal cortex, and the ventromedial cortex (Barrett et al., 2007).

1.3 Neural Basis of Emotional Contagion

Hatfield et al. (1994) define emotional contagion as the "tendency to automatically mimic and synchronize the facial expressions, vocalizations, postures, and movements with those of another person and consequently to converge emotionally" (pp. 153–154). Emotional contagion is the transfer of emotions among group members (Barsade, 2002) and organizations are arenas where such transfer of emotions can occur on a regular basis. Neuroscience could help us understand why some types of emotions may "spread" among individuals interacting with one another on a regular basis, as in the case of employees working in the same organization. Specifically, the mirror neuron system (Iacoboni, 2009), described in Chapters 6 and 8, can help us understand why people who frequently interact with one another can share similar emotions.

Barrett et al. (2007) found that the observation of an emotion activates the neural representation of that emotion. They conducted an fMRI study in which participants inhaled odorants producing a strong feeling of disgust and observed video clips showing the emotional facial expression of disgust. Observing such faces and feeling disgust activated the same sites of the anterior insula and to some extent the anterior cingulate cortex (Wicker et al., 2003). Öhman (2005) found that masked facial stimuli activate the amygdala, as do masked pictures of threatening animals such as snakes and spiders. When the stimulus conditions allow conscious processing, the amygdala response to feared stimuli is enhanced and a cortical network that includes the anterior cingulate cortex and the anterior insula is activated. However, the initial amygdala response to a fear-relevant but non-feared stimulus (e.g., pictures of spiders for a snake phobic) disappears with conscious processing and the cortical network is not recruited. Instead, there is activation of the dorsolateral and orbitofrontal cortices that appears to inhibit the amygdala response.

1.4 Somatic Marker Hypothesis and Emotional Regulation

Emotional stimuli elicit bodily changes (Damasio, 1994). The somatic marker hypothesis (Damasio, 1994; Bechara et al., 2000; Naqvi, Shiv, and Bechara, 2006; Deak, 2011) suggests that decisions are helped by emotions in the form of bodily states that are elicited during the deliberation of future consequences and that mark different options for behavior as being advantageous or disadvantageous. As Naqvi et al. (2006, p. 260) contend, "decision making involves not only the cold-hearted calculation of expected utility based upon explicit knowledge of outcomes but also subtler and sometimes covert processes that depend critically upon emotion." "The somatic marker hypothesis proposes that individuals make judgements not only by assessing the severity of outcomes and their probability of occurrence, but also and primarily in terms of emotional quality" (Bechara et al., 2000, p. 305). It posits that "emotional states are triggered in the brain before and after decision making" (Lee and Chamberlain, 2007, p. 34).

The main assumption of this hypothesis is that decision making is a process that is influenced by marker signals that arise in bioregulatory processes, including those that express themselves in emotions and feelings. "The fundamental notion of the somatic marker hypothesis is that bio-regulatory signals, including those that constitute feeling and emotion, provide the principal guide for decisions and are the basis for the development of the 'as-if body loop' mode of operation" (Bechara et al., 2000, p. 306). The somatic marker hypothesis provides a systems-level neuroanatomical and cognitive framework for decision making and is influenced by emotion. This influence can occur at multiple levels of operation, some of which occur consciously, and some of which occur non-consciously (Bechara and Damasio, 2005).

In a study using the Gambling Task, Bechara et al. (2000) found that patients with amygdala damage as well as those with ventromedial prefrontal cortex damage were impaired and unable to develop anticipatory skin conductance responses (SCRs) while they pondered risky choices. However, patients with ventromedial prefrontal cortex damage were able to generate SCRs when they received a reward or a punishment, unlike patients with amygdala damage. According to Naqvi et al. (2006), before a decision is made, a somatic state occurs when one contemplates the outcomes of choices. After a choice is made, the somatic state that results from the outcome (either reward or punishment) works to form learned associations between choice and outcome. Naqvi et al. (2006) found that the amygdala plays an important role in providing these learned associations between decision outcomes and choices. The "hot stove effect" is an illustrative example. Touching a hot stove can result in burning one's finger, which can create an emotional response – fear of a hot stove. This learned association between a hot stove and being burned (the outcome of touching the hot stove) leads us to avoid touching a hot stove in the future.

1.5 Emotions and Empathy

Decety and Jackson (2006, p. 54), defined empathy as the "capacity to understand and respond to the unique affective experiences of another person." They also contend that empathy has three primary components: (1) an affective response to another person, which often, but not always, entails sharing that person's emotional state; (2) a cognitive capacity to take the perspective of the other person; and (3) emotion regulation. Decety and Jackson (2006) suggest that other components, including people's ability to monitor and regulate cognitive and emotional processes to prevent confusion between self and other, are equally necessary parts of a functional model of empathy. Brain structures such as the insula, the anterior cingulate cortex, and the right temporoparietal region are implicated in empathy.

Empathy is a critical aspect of human emotion that influences the behavior of individuals as well as the functioning of society. Rameson, Morelli, and Lieberman (2011) found that across conditions, higher levels of self-reported experienced empathy were associated with greater activity in the medial prefrontal cortex (mPFC). High trait empathy participants displayed greater experienced empathy and stronger mPFC responses than low trait empathy individuals under cognitive load, suggesting that empathy is more automatic for individuals high in trait empathy. Activity in the mPFC was also correlated with daily helping behavior. Self-report of empathic experience and activity in empathy-related areas, notably the mPFC, were higher in the empathize condition than in the load condition, suggesting that empathy is not a fully automatic experience.

According to Oatley and Johnson-Laird (1987) the organism has two operative systems with which to face an unpredictable environment: (1) the emotional system, considered as an emergency system, able to interrupt the ongoing activity to rapidly select a new operative scheme; and (2) the cognitive system, considered a more complex and evolved system, but requiring more time to carry out its work. In their model, the authors argue that emotions provide a biological solution to certain problems of transition between plans, in systems with multiple goals. The function of emotions is to accomplish and maintain these transitions, and to communicate them to ourselves and others. Transitions occur at significant junctures of plans when the evaluation of success in a plan changes. Complex emotions are derived from a small number of basic emotions and arise at junctures of social plans.

If neuroscience can contribute to a better understanding of emotional labor, it can also provide useful guidelines for improving emotional intelligence. Emotional intelligence refers to a set of competencies that are essential features of human social life (Goleman, 2006; Krueger et al., 2009). It includes the ability to understand and control one's own emotions as well as those of others (Goleman, 2006). To the extent that the research described above concludes that several brain structures play an important role in emotional reactions, emotional regulation and emotional contagion, one may speculate that an understanding of neuroscience could provide useful insights into emotional intelligence. It could also contribute to the development of training methods aimed at helping employees and managers to become emotionally competent.

Krueger et al. (2009) studied a unique sample of combat veterans from the Vietnam Head Injury Study, which is a prospective, long-term follow-up study of veterans with focal penetrating head injuries. They administered the Mayer-Salovey-Caruso Emotional Intelligence Test to examine strategic emotional intelligence (the competency to understand emotional information and to apply it to the management of the self and of others) and experiential emotional intelligence (the competency to perceive emotional information and to apply it to integration into thinking). Krueger et al. (2009) found that the two competencies underlying emotional intelligence depend on distinct neural prefrontal cortex substrates. First, they noticed that damage to the ventromedial prefrontal cortex diminished strategic emotional intelligence, and therefore hinders the understanding and managing of emotional information. Second, they found that damage to the dorsolateral prefrontal cortex diminishes experiential emotional intelligence, and therefore hinders the perception and integration of emotional information. These findings indicate a neural basis for emotional intelligence.

2 NEURAL BASIS OF UNCONSCIOUS BIAS

Although psychologists are studying the concept of unconscious or implicit bias, it has yet to gain traction in organizational behavior and management. Because of greater diversity efforts and organizational behavior research on diversity in the workplace, it would seem important to pay more attention to issues of implicit (or unconscious bias) in these fields.

People may sometimes lack knowledge of and control over the causes and consequences of their own actions. What is meant by unconscious bias? When is a bias unconscious? "Unconscious bias refers to individuals' lack of awareness of the effects of their own actions on other people or social institutions" (Blanton and Jaccard, 2008, p. 279). For example, people may hold stereotypes about others because of their race or religion or other characteristics. Such preconceived ideas are often unconscious although they may affect actual behavior. Hence, implicit bias is difficult to detect and therefore to reduce in the workplace.

However, knowing that our biases can be unconscious could facilitate self-examination and self-scrutinization of the decisions we make regarding those who are dissimilar to us.

The study of the neural foundations of unconscious bias is relevant to a discussion on the neural basis of emotions in organizations. As Phelps et al. (2014) suggest, "in our everyday lives, the stimuli most likely to elicit emotional responses are other people." The likelihood that these other people induce emotional reactions can be exacerbated when they are perceived as out-group members.

Social categorization theory (Tajfel, 1970; Tajfel and Turner, 1979) indicates that people have a tendency to categorize others along dimensions such as age, gender, race, ethnicity, and the like. In doing so, they tend to attribute certain characteristics to members of the groups so formed and assume that people belonging to the same group have similar characteristics and those belonging to different groups have different characteristics. Such social categorization may form the basis of prejudice and discrimination in life. In fact, merely dividing people into groups is enough to trigger discrimination (Tajfel, 1970).

According to Blanton and Jaccard (2008), although it is feasible to argue that people lack knowledge of both the causes and the consequences of their actions, this alone does not provide a basis for claiming that people possess racist attitudes that escape subjective awareness. Thus, understanding the neural basis of people's reactions to dissimilar others could provide useful insights for reducing biases in the workplace. For example, Sessa et al. (2012, p. 315) note that "a complete understanding of racial biases cannot be achieved without an in-depth examination of how racial prejudice influences basic cognitive processes such as those involved in face representation."

Life in organizations occurs in the context of social interactions. Social interactions also require the ability to take into account social norms and mentalize (or have a theory of mind, that is, the ability to infer the mental states of others). Mentalizing involves integrating the perspectives of others (Güroğlu, Van den Bos, and Crone, 2009). The importance of social interactions cannot be underestimated because "human survival depends in large part on the formation of alliances and accurate judgments" (Cacioppo, 2002, p. 820). It is well understood that human beings are a social species, so much so that the survival and success of our species depend on its ability to function in complex social interactions (Gallese, Keysers, and Rizzolatti, 2004).

Psychologists have long contended that people exhibit in-group and out-group biases (Tajfel, 1970; Tajfel and Turner, 1979). This process can be conscious but often it is unconscious and automatic. In fact, people exhibit a natural tendency to associate with those who are similar to themselves and maintain some type of social distance with those who are dissimilar. To some extent, this natural tendency helps satisfy certain basic psychological needs in people who cooperate with in-group members (Allport, 1954).

2.1 Neuroscience, Unconscious Bias, and Diversity

In an era of diversity and multiculturalism, organizational scholars would better be served by studying the effects of unconscious bias on behavior in organizations. Studying the neural foundations of unconscious bias could contribute to the refinement of theories on diversity and its consequences in organizations. After all, organizations are arenas where people interact with others. As a result, they are environments where unconscious bias could play out on a daily basis. Hence, relying on neuroscience to uncover the determinants of unconscious bias is important because "the neuroscience literature suggests that implicit processing always plays a role in determining our emotions, attitudes, and behavior" (Becker and Menges, 2013, p. 220).

The dual system (discussed in earlier chapters as System 1 and System 2) could help assess the importance of neuroscience in understanding unconscious bias. An implicit attitude (related to System 1) is rapid, automatic, and comprises unconscious evaluations in response to stimuli, whereas an explicit attitude (related to System 2) is a relatively slower, deliberative, and conscious evaluation based on contextual information. One may hold negative attitudes toward others without overtly expressing them. An expressed attitude is one that people report (Cunningham et al., 2007). The neural study of unconscious bias is important because it can provide the opportunity to understand personal biases in the workplace and reduce them. As Cropanzano and Becker (2013, p. 307) note, "neuroscientific methods provide the most reliable window into the nonconscious brain that is currently available."

2.2 Neuroscience of Unconscious Bias

There are several characteristics that differentiate individuals from one another. For example, diversity researchers have identified elements of surface diversity (race, gender, age) and deep diversity (education, values, personality) that differentiate people. Of these, the most visible, pervasive, and often controversial is race. Indeed, racial attitudes influence people's behavior in all parts of the world. Although attempts to explain the causes of negative racial attitudes have been made by sociologists, it is unclear why they still persist. It is thought that fearing something or someone different could be traced to our evolutionary roots. Perhaps, to survive, early humans developed adaptive measures such as avoiding anything perceived as different and therefore threatening.

2.2.1 Neural basis of race-related unconscious bias

According to Malpass and Kravitz (1969), people tend to better recognize faces of their own race than faces of other races, a phenomenon known as the own-race-bias (ORB). Studies on the recognition of faces of an ethnic group different from one's own, reveal a robust recognition deficit for faces of the respective out-group (cross-race effect or own-race bias) and a tendency to respond less cautiously with respect to out-group faces (Sporer, 2001). In fact, "humans are better at recognizing individuals of their own race than of other races" (Phelps, 2001, p. 775). Racial prejudice influences the precision of visual representations of other-race faces in visual working memory (Sessa et al., 2012). This implies that the less people are exposed to others of a different race, the less likely they are to differentiate their faces. In other words, the lack of familiarity with people from a different race may contribute to own-race-bias.

The familiarity effect may reduce the negative perceptions that people may have of dissimilar others. For example, Phelps et al. (2000) found activation in the amygdala in white participants who tended to automatically perceive unfamiliar black faces as potentially threatening and fear relevant. However, in a subsequent study, Phelps, Cannistraci, and Cunningham (2003) found that when participants became familiar with out-group members, they experienced less activation of the amygdala. Hart, Whalen, and Shin (2000) studied the neural reactions to faces in one's own race and a person from a different race using a sample of black and white participants. The results showed greater activation of the amygdala when participants view out-group faces compared to in-group faces, but at a later phase. However, initial presentations did not show a significant difference in amygdala activation between in-group and out-group faces. Amodio et al. (2004) used electroencephalography (EEG) to examine unintentional race-biased responses and found that responses attributed to race bias produced greater error-related negativity (ERN) than responses not attributed to bias. These findings indicate that race-bias responses may be made despite the activation of neural systems designed to detect bias and to recruit controlled processing.

Mirror neurons could help explicate discriminatory behavior. To some extent, they could represent the deep causes of discriminatory behavior because they reside in the "deep-brain structures of the limbic system" (Becker, Cropanzano, and Sanfey, 2011, p. 941). In the mirror neuron system, "the observation of an action leads to the activation of parts of the same cortical neural network that is active during its execution" (Gallese et al., 2004, p. 396). Another reason could be that the mirror neuron system may lead people to imitate the attitudes and emotions of others. For example, perceiving that in-group members display negative emotions and attitudes toward out-group members may lead an in-group member to act likewise. Hence, discriminatory attitudes and behaviors could be neurologically contagious.

Phelps et al. (2000) conducted two experiments designed to explore the neural basis of racial attitudes. In the first experiment they found that the strength of amygdala activation to black versus white faces was correlated with two indirect (unconscious) measures of race evaluation (Implicit Association Test, IAT and potentiated startle test) but not with the direct (conscious) expression of race attitudes. In the second experiment, these patterns were not obtained when the stimulus faces belonged to familiar and positively regarded black and white individuals. The authors conclude that amygdala and behavioral responses to black versus white faces in white subjects reflect cultural evaluations of social groups modified by individual experience. They also found that hemodynamic activation at the level of the amygdala is modulated by implicit prejudice toward members of a different race.

These findings bear two important implications for diversity in the workplace: familiarity with different others and whether the out-group member is well perceived. To reduce race-related unconscious bias, organizations may provide opportunities for interactions with employees from various racial and ethnic backgrounds. Such frequent interactions may diminish own-race-bias. Promoting minorities and highlighting their accomplishments when needed could lead to a cognitive re-evaluation of these minorities but also their racial or ethnic group. Such strategies could contribute to diminishing negative reactions, as illustrated by activation in the amygdala. Being exposed to people from different races in the workplace can influence a person's attitude toward other races in their social life.

Cunningham et al. (2004) studied participants' reactions to black and white faces and found that when the faces were presented for 30 milliseconds, activation in the amygdala was greater for black than for white faces. When the faces were presented for 525 milliseconds, this difference was significantly reduced. Regions of the frontal cortex associated with control and regulation showed greater activation for black and white faces. The difference in amygdala activation was stronger when participants expressed more racial bias on the Implicit Association Test (IAT). These results suggest that "implicit negative associations to a social group may result in an automatic emotional response when encountering members of that group" (Cunningham et al., 2004, p. 811). Sessa et al. (2012) studied the neural basis of unconscious prejudice using electroencephalography. They found that high-prejudiced participants encoded black people's faces with a lower degree of precision compared to low-prejudiced participants. These findings indicate that the class of mental operations affected by implicit racial prejudice includes basic cognitive mechanisms underpinning the encoding and maintenance of the visual representation of faces in visual working memory.

Krendl, Kensinger, and Ambady (2012) studied participants' regulation of negative bias to stigma. Prior to the neural analysis, they had the participants take an IAT to measure their attitudes toward homelessness and alcoholism. They found that participants had higher activity in the anterior cingulate cortex (ACC) and the lateral and medial prefrontal cortex (PFC) during their initial attempt to regulate their negative affect to images of stigmatized individuals as compared to non-stigmatized individuals.

Krendl et al. (2012) conclude that stigma regulation may be a more immediate response, whereas emotion regulation may be more prolonged. They also found that during the implicit task, PFC regions were more active to the more unpleasant targets. "When participants were asked to make explicit judgments of stigmatized targets, they engage the VLPFC and the ACC, perhaps to automatically regulate their responses" (Krendl et al., 2012, p. 724). The authors also found a correlation between activation in some of the PFC regions during stigma regulation and participants' implicit bias. In a study focusing on the neural basis of error detection concerning people from different races, Amodio et al. (2004) found that the ACC was more active when participants had the potential for making race-based errors even before those errors occurred.

Richeson et al. (2003) had white participants complete an unobtrusive measure of racial bias, then interact with a black individual, and finally complete an unrelated Stroop color-naming test. In a separate fMRI session, participants were presented with unfamiliar black male faces, and the activity of brain regions thought to be critical to executive control was assessed. They found that racial bias predicted activity in right dorsolateral prefrontal cortex in response to black faces. They also found that activity in this region predicted Stroop interference after an actual interracial interaction, and it statistically mediated the relation between racial bias and Stroop interference. These results are consistent with a resource-depletion account of the temporary executive dysfunction seen in racially biased individuals after interracial contact.

Ito and Urland (2003) investigated the degree to which perceivers automatically attend to and encode social category information. Event-related brain potentials were used to assess attentional and working-memory processes online as participants were presented with pictures of black and white males and females. The authors found that attention was preferentially directed to black targets very early in processing (by about 100 milliseconds after stimulus onset) in both experiments. Attention to gender also emerged early but occurred about 50 milliseconds later than attention to race. These working-memory processes were sensitive to both the explicit categorization task participants were performing as well as more implicit, task-irrelevant categorization dimensions. These results are consistent with models suggesting that information about certain category dimensions is encoded relatively automatically.

Van Bavel, Packer, and Cunningham (2008) randomly assigned participants to a mixed-race team and used fMRI to identify brain regions involved in processing novel in-group and out-group members independently of preexisting attitudes, stereotypes, or familiarity. Whereas previous research on intergroup perception found amygdala activity, typically interpreted as negativity, in response to stigmatized social groups, the authors found greater activity in the amygdala, fusiform gyri, orbitofrontal cortex, and dorsal striatum when participants viewed novel in-group faces than when they viewed novel out-group faces. They also found that activity in orbitofrontal cortex mediated the in-group bias in self-reported liking for the faces. These in-group biases in neural activity were not moderated by race or by whether participants explicitly attended to team membership or race, a finding suggesting that they may occur automatically.

Golby et al. (2001) investigated the neural substrates of same-race memory superiority using fMRI. European-American (EA) and African-American (AA) males underwent fMRI while they viewed photographs of AA males, EA males and objects under intentional encoding conditions. They found that recognition memory was superior for same-race versus other-race faces. Individually defined areas in the fusiform region that responded preferentially to faces had greater response to same-race versus other-race faces. Across both groups, memory differences between same-race and other-race faces correlated with activation in left fusiform cortex and right parahippocampal and hippocampal areas. These results suggest that differential activation in fusiform regions contributes to same-race memory superiority. Participants that exhibited the strongest effect also displayed the greatest own-race-bias.

2.2.2 Unconscious bias and mentalizing

Mitchell, Macrae, and Banaji (2006) explored mentalizing based on whether the other individual is similar or dissimilar. They found that mentalizing about a similar other engaged a region of the ventral medial prefrontal cortex linked to self-regulation thought, whereas mentalizing about a dissimilar other engaged a more dorsal subregion of the medial prefrontal cortex. "Mentalizing on the basis of self-knowledge can only take place if another person's internal experience is assumed to be comparable to one's own" (Mitchell et al., 2006, p. 656). This is particularly important in organizations where people from various ethnic, racial, and gender backgrounds coexist.

Using electroencephalographic oscillations as an index of perception–action coupling, Gutsell and Inzlicht (2010) found that participants displayed activity over the motor cortex when acting and when observing in-groups act, but not when observing out-groups act. "When people connect with others, they resonate with them by adopting their postures, intonations, and facial expressions, but also their motivational states and emotions" (Gutsell and Inzlicht, 2010, p. 841). These findings provide evidence from brain activity that a spontaneous and implicit simulation of others' action states may be limited to close others and, without active effort, may not be available for out-groups. People are less likely to mentally simulate the actions of others when these others are out-group members. This represents a challenge that must be overcome for diverse teams to be effective. To be effective, teams must have some shared mental models. Organizational scholars could explore whether diverse teams have less shared mental models than homogeneous teams.

2.3 Empathy and the Reduction of Unconscious Bias

Showing empathy can help reduce unconscious bias and improve the effectiveness of diversity initiatives. Research on diverse teams reveals a double-edged sword (Rypma et al., 2006). On one side, diversity makes more information available and encourages creativity. On the other side, diversity can reduce cohesion and information sharing. Roberson, Holmes, and Perry (2017, p. 212) echo this observation and note that "despite a growing body of literature on diversity and firm performance, our review of research across fields, theoretical traditions, and levels of analysis suggests that the relationship is not a simple one." Hence, the complexity of the concept could lead us to suggest that diversity can be helpful or destructive depending on how it is managed. To overcome the potential negative impact of diversity, it is important to consider the role of empathy (discussed earlier in this chapter).

Empathy refers to the ability to comprehend and vicariously share the feelings and thoughts of other people, according to the perception-action model (De Vignemont and Singer, 2006). Empathy acts as a key motivator in helping attitude and cooperative behavior. Preston and De Waal (2002) developed a perception-action model of empathy, which holds that the observation or imagination of another person in a particular emotional state or action automatically activates a representation of that same state/action in the observer, including its associated autonomic and somatic responses. However, people tend to show more empathy for in-group members than for out-group members.

Recent studies have shown that perceiving the pain of others activates brain regions in the observer associated with both somatosensory and affective-motivational aspects of pain, principally involving regions of the anterior cingulate cortex and the anterior insula (Contreras-Huerta, Baker, and Reynolds, 2013). Racial bias modulates the degree of these empathic neural responses such that stronger neural activation is elicited by observing pain in people of the same racial group compared to people of another racial group. Contreras-Huerta et al. (2013) observed that participants were more sensitive to pain induced in members of their own-racial group than in members of the other racial group. Consequently, the authors suggested that race may be an automatic and unconscious mechanism that drives the initial neural responses to observed pain in others.

De Vignemont and Singer (2006) propose two major roles for empathy: an epistemological role and a social role. In its epistemological role, empathy provides information about the future actions of other people, and important environmental properties. In its social role, empathy serves as the origin of the motivation for cooperative and prosocial behavior, as well as help for effective social communication. According to De Vignemont and Singer (2006), "there is empathy if: (i) one is in an affective state; (ii) this state is isomorphic to another person's affective state; (iii) this state is elicited by the observation or imagination of another person's affective state; (iv) one knows that the other person is the source of one's own affective state" (p. 435).

Shared neural representations, self-awareness, mental flexibility, and emotion regulation constitute the basic macro-components of empathy, which are underpinned by specific neural systems (Decety and Jackson, 2006). These shared neural representations could help improve the effectiveness of diverse teams. However, it is important to understand whether shared neural representations could exist in racially and ethnically diverse teams. Organizational scholars could address this question by comparing racially homogeneous and heterogeneous groups.

Showing empathy could prevent people from dehumanizing out-group members (Bandura, Underwood, and Fromson, 1975; Jack, Dawson, and Norr, 2013) or being morally disengaged (Bandura, 1999) when considering the plight of others who are dissimilar in several attributes. Dehumanization refers to the process of seeing others as less human (Jack et al., 2013). For Jack et al. (2013), dehumanization is linked to a neural constraint on cognition, which makes it difficult for someone to be both empathetic and analytic. Harris and Fiske (2006) studied the neural basis of dehumanization and observed a decreased activation in the ventromedial prefrontal cortex when participants viewed images of out-group members who were also described as less warm and incompetent. Diversity initiatives in organizations can be successful when people show empathy and refrain from dehumanizing others. Diversity in organizations implies the "coexistence" of individuals who may perceive some members as in-group members and others as out-group members. Hence, the temptation for potential dehumanization of those who are perceived as out-group members should not be neglected.

Conclusion

In this book, I have discussed the neural basis of organizational behavior. In so doing, I have highlighted the neural basis of topics such as decision making, creativity and innovation, motivation, emotions, ethics, trust and cooperation, and unconscious bias in the workplace. Although I did not devote specific chapters to topics such as organizational change, organizational culture, and personality, the treatment of the other topics has alluded to them. Organizational neuroscience is construed as a multidisciplinary topic drawing from neuroscience and other social sciences that use neuroscientific methods. For example, organizational neuroscience draws from evolutionary psychology to understand human behavior in organizations. Evolutionary psychology emphasizes the fact that human beings are hard-wired to act in certain ways and "holds that although human beings today inhabit a thoroughly modern world of space exploration and virtual realities, they do so with the ingrained mentality of Stone Age hunter-gatherers" (Nicholson, 1998, p. 135). In fact, some of the functions of the brain specialize over time to respond to environmental constraints. As Nicholson (1998) argued, "you can take the person out of the Stone Age but you cannot take the Stone Age out of the person" (ibid.).

By exploiting this type of knowledge about brain organization and function, and determining which brain systems are associated with a particular behavior, researchers may be able to better understand the processes driving the behavior in question (Cohen, 2005). Moreover, "neural mechanisms are largely homogenous across all individuals and are recruited to respond to numerous different organizational situations. That is to say, each neuron operates in the same way, and all brains are organized in a similar fashion" (Becker, Cropanzano, and Sanfey, 2011, p. 936).

To gain legitimacy in the organizational sciences, organizational neuroscience must not follow the early path of neuroscience where early studies were only descriptive, exploratory, and often atheoretical (Rick, 2011). Thus, organizational neuroscientists should develop theories based on findings from neuroscience and organizational science. They can also draw from research in allied disciplines such as neuroeconomics, social cognitive neuroscience, cognitive neuroscience, and cognitive psychology. Other disciplines, such as neurotheology, neuromarketing, neurophilosophy, and other social sciences that use neuroscientific methods could provide insights to research in organizational neuroscience. For example, employees interact with customers when selling their companies' products or explaining the features of certain products. In addition, organizations are adopting the view that employees are internal customers. Hence, employees consider one another as customers when their tasks are interrelated. Likewise, knowledge from the nascent field of neurotheology could help enhance our understanding of emerging concepts in the workplace such as mindfulness and spirituality.

A key question for organizational neuroscience scholars is whether to focus only on neuroscientific studies that involve the brain or to integrate other biological studies that do not involve activation of brain structures. For example, should genetic studies of organizational behavior be part of organizational neuroscience? Similarly, should studies using other biological methods be under the purview of organizational neuroscience? Answers to these questions are important because they would help delineate the field of organizational neuroscience. It is also important to determine whether the study of organizational neuroscience should broaden itself to include the neural foundations of entrepreneurship (neuroentrepreneurship), strategic management (neurostrategy), or management information systems.

Although this book has not covered the neural basis of all topics discussed in the organizational sciences, it has provided a valuable account of research on the link between the human brain and actual behaviors displayed by managers and employees. In this regard, it will contribute to the establishment of organizational neuroscience as a legitimate field of study.

References

Abraham, A., Pieritz, K., and Thybusch, K. et al. (2012). Creativity and the brain: Uncovering the neural signature of conceptual expansion. *Neuropsychologia, 50*(8), 1906–1917.

Abraham, A., and Windmann, S. (2007). Creative cognition: The diverse operations and the prospect of applying a cognitive neuroscience perspective. *Methods, 42*(1), 38–48.

Abraham, R. (1998). Emotional dissonance in organizations: Antecedents, consequences, and moderators. *Genetic, Social, and General Psychology Monographs, 124*(2), 229–246.

Adams, J.S. (1965) Inequity in social exchange. In L. Berkowitz (ed.), *Advances in Experimental Social Psychology* (Vol. 2, pp. 267–299). New York: Academic Press.

Adolphs, R. (2001). The neurobiology of social cognition. *Current Opinion in Neurobiology, 11*(2), 231–239.

Adolphs, R. (2003). Cognitive neuroscience of human social behavior. *Nature Reviews Neuroscience, 4*, 165–178.

Ahlfors, S.P., and Mody, M. (2016). Overview of MEG. *Organizational Research Methods.* Doi.org/10.1177/1094428116676344.

Akinola, M. (2010). Measuring the pulse of an organization: Integrating physiological measures into the organizational scholar's toolbox. *Research in Organizational Behavior, 30*, 203–223.

Albrecht, K., Abeler, J., Weber, B., and Falk A. (2014). The brain correlates of the effects of monetary and verbal rewards on intrinsic motivation. *Frontiers of Neuroscience, 8*(303), 1–10.

Alderfer, C.A. (1969). An empirical test of a new theory of human needs. *Organizational Behavior and Human Performance, 4*(2), 142–175.

Alderfer, C.A. (1972). *Human Needs in Organizational Settings.* New York: The Free Press of Glencoe.

Alexiou, K., Zamenopoulos, T., Johnson, J.H., and Gilbert, S.J. (2009). Exploring the neurological basis of design cognition using brain imaging: Some preliminary results. *Design Studies, 30*(6), 623–647.

Alexopoulos, J., Pfabigan, D.M., and Lamm, C. et al. (2012). Do we care about the powerless third? An ERP study of the three-person ultimatum game. *Frontiers in Human Neuroscience, 6*, 1–9.

Allen, A.P., and Thomas, K.E. (2011). A dual process account of creative thinking. *Creativity Research Journal, 23*(2), 109–118.

Allison, J. (1983). *Behavioral Economics*. New York: Praeger.

Allport, G.W. (1954). *The Nature of Prejudice*. Reading, MA: Addison Wesley.

Amabile, T.M. (1996). *Creativity in Context: Update to the Social Psychology of Creativity*. Boulder, CO: Westview Press.

Amodio, D.M., Harmon-Jones, E., and Devine, P.G. et al. (2004). Neural signals for the detection of unintentional race bias. *Psychological Science, 15*(2), 88–93.

Andrews-Hanna, J.R., Smallwood, J., and Spreng, N.R. (2014). The default network and self-generated thought: Component processes, dynamic control, and clinical relevance. *Annals of the New York Academy of Sciences, 1316*, 29–52.

Ansoff, H. (1965). *Corporate Strategy*. New York: McGraw Hill.

Anticevic, A., Cole, M.W., and Murray, J.D. et al. (2012). The role of default network deactivation in cognition and disease. *Trends in Cognitive Sciences, 16*(12), 584–592.

Arrow, K.J. (1974). Limited knowledge and economic analysis. *American Economic Review, 64*(1), 1–10.

Ashkanasy, N.M. (2013). Neuroscience and leadership: Take care not to throw the baby out with the bathwater. *Journal of Management Inquiry, 22*(3), 311–313.

Ashkanasy, N.M., Becker, W.J., and Waldman, D.A. (2014). Neuroscience and organizational behavior: Avoiding both neuro-euphoria and neuro-phobia. *Journal of Organizational Behavior, 35*(7), 909–919.

Aston-Jones, G., and Cohen, J.D. (2005). An integrative theory of locus coeruleus-norepinephrine function: Adaptive gain and optimal performance. *Annual Review of Neuroscience, 28*, 403–450.

Avolio, B.J., Bass, B.M., and Jung, D.I. (1999). Re-examining the components of transformational and transactional leadership using the multifactor leadership questionnaire. *Journal of Occupational and Organizational Psychology, 72*(4), 441–462.

Axelrod, R., and Hamilton, W.D. (1981). The evolution of cooperation. *Science, 211*(4489), 1390–1396.

Aziz–Zadeh, I., Liew, S.L., and Dandekar, F. (2013). Exploring the neural correlates of visual creativity. *Social Cognitive and Affective Neuroscience, 8*(4), 475–480.

Balleine, B.W. (2007). The neural basis of choice and decision making. *Journal of Neuroscience, 27*(31), 8159–8160.

Balthazard, P.A., and Thatcher, R.W. (2015), Neuroimaging modalities and brain technologies in the context of organizational neuroscience. In D.A. Waldman and P.A. Balthazard (eds), *Organizational Neuroscience (Monographs in Leadership and Management* (Vol. 7, pp. 83–113). Bingley, UK: Emerald Group Publishing Limited.

Balthazard, P.A., Waldman, D.A., Thatcher, R.W., and Hannah, S.T. (2012). Differentiating transformational and non-transformational leaders on the basis of neurological imaging. *The Leadership Quarterly*, *23*(2), 244–258.

Bandura, A. (1999). Moral disengagement in the perpetration of inhumanities. *Personality and Social Psychology Review*, *3*(3), 193–209.

Bandura, A., Underwood, B., and Fromson, M.E. (1975). Disinhibition of aggression through diffusion of responsibility and dehumanization of victims. *Journal of Research in Personality*, *9*(4), 253–269.

Barker, A.T., Jalinous, R., and Freeston, I.L. (1985). Non-invasive magnetic stimulation of human motor cortex. *Lancet*, *1*(8437), 1106–1107.

Barnett, T. (2001). Dimensions of moral intensity and ethical decision making: An empirical study. *Journal of Applied Social Psychology*, *31*(5), 1038–1057.

Baron, R.A. (2006). Opportunity recognition as pattern recognition: How entrepreneurs connect the dots to identify new business opportunities. *Academy of Management Perspectives*, *20*(1), 104–119.

Barrett, L.F., Mesquita, B., Ochsner, K.N., and Gross, J.J. (2007). The experience of emotion. *Annual Review of Psychology*, *58*, 373–403.

Barsade, S.G. (2002). The ripple effect: Emotional contagion, and its influence on group behavior. *Administrative Science Quarterly*, *47*(4), 644–675.

Bartz, J.A., Zaki, J., Bolger, N., and Ochsner, K.N. (2011). Social effects of oxytocin in humans: Context and person matter. *Trends in Cognitive Sciences*, *15*(7), 301–309.

Bass, B.M. (1985). *Leadership and Performance Beyond Expectations*: New York: Free Press/London: Collier Macmillan.

Bass, B.M., and Avolio, B.J. (1995). *Multifactor Leadership Questionnaire*, Redwood City, CA: MindGarden.

Baumgartner, T., Heinrichs, M., and Vonlanthen, A. et al. (2008). Oxytocin shapes the neural circuitry of trust and trust adaptation in humans. *Neuron*, *58*(4), 639–650.

Baumgartner, T., Knoch, D., and Hotz, P. et al. (2011). Dorsolateral and ventromedial prefrontal cortex orchestrate normative choice. *Nature Neuroscience*, *14*(11), 1468–1476.

Beaty, R.E., Benedek, M., Silvia, P.J., and Schacter, D.L. (2016). Creative cognition and brain work dynamics. *Trends in Cognitive Sciences*, *20*(2), 87–95.

Beaty, R.E., Silvia, P.J., and Nusbaum, E.C. et al. (2014). The roles of associative and executive processes in creative cognition. *Memory and Cognition*, *42*(7), 1186–1197.

Bechara, A., and Damasio, H. (2005). The somatic marker hypothesis. A neural theory of economic decisions. *Games and Economic Behavior*, *52*(2), 336–372.

Bechara, A., Damasio, H., and Damasio, A.R. (2000). Emotion, decision making and the orbitofrontal cortex. *Cerebral Cortex*, *10*(3), 295–307.

Bechtel, W. (2002). Aligning multiple research techniques in cognitive neuroscience: Why is it important? *Philosophy of Science*, *69*(S3), S48–S58.

Becker, W.J., and Cropanzano, R. (2010). Organizational neuroscience: The promise and prospects of an emerging field. *Journal of Organizational Behavior*, *31*(7), 1055–1059.

Becker, W.J., and Menges, J.I. (2013). Biological implicit measures in HRM and OB: A question of how not if. *Human Resource Management Review*, *23*(3), 219–228.

Becker, W.J., Cropanzano, R., and Sanfey, A.G. (2011). Organizational neuroscience: Taking organizational theory inside the neural black box. *Journal of Management*, *37*(4), 933–961.

Beeler, J.A., Daw, N.D., Frazier, C.R.M., and Zhuang, X. (2010). Tonic dopamine modulates exploitation of reward learning. *Frontiers in Behavioral Neuroscience*. Doi.org/10.3389/fnbeh.2010.00170.

Behrens, T.E.J., Woolrich, M.W., Walton, M.E., and Rushworth, M.F.S. (2007). Learning the value of information in an uncertain world. *Nature Neuroscience*, *10*(9), 1214–1221.

Benkler, Y. (2011a). *The Penguin and the Leviathan: How Cooperation Triumphs Over Self-interest*. New York: Crown Business.

Benkler, Y. (2011b). The unselfish gene. *Harvard Business Review*, July–August, 3–11.

Berg, J., Dickhaut, J., and McCabe, K. (1995). Trust, reciprocity, and social history. *Games and Economic Behavior*, *10*(1), 122–142.

Berger, C.R. (1979). Beyond initial understanding: Uncertainty, understanding, and the development of interpersonal relationships. In H. Giles and R.N. St. Clair (eds), *Language and Social Psychology* (pp. 122–144). Oxford: Blackwell.

Berger, H. (1929). Uber das electrenkephalogramm des menschen. *Archiv für Psychiatrie und Nervenkrankheiten*, *87*(1), 527–570.

Bernoulli, D. ([1738] 1954). Exposition of a new theory on the measurement of risk. *Econometrica*, *22*(1), 23–36.

Berridge, K.C. (2004). Motivational concepts in behavioral neuroscience. *Physiology and Behavior*, *81*(2), 179–209.

Bethlehem, R.A.I., Van Honk, J., Auyeung, B., and Baron-Cohen, S. (2013). Oxytocin, brain physiology, and functional connectivity: A review of intranasal oxytocin fMRI studies. *Psychoneuroendocrinology*, *38*(7), 962–974.

Beugré, C.D. (2009). Exploring the neural foundations of organizational justice: A neurocognitive model. *Organizational Behavior and Human Decision Processes*, *110*(2), 129–139.

Beugré, C.D. (2010). Brain and human behavior in organizations: A field of neuro-organizational behavior. In A.A. Stanton, M. Day, and I. Welpe (eds), *Neuroeconomics and the Firm* (pp. 289–303). Cheltenham, UK and Northampton, MA, USA: Edward Elgar Publishing.

Beugré, C.D. (2016). A neurocognitive model of entrepreneurial cognitions. *Current Topics in Management*, *18*, 17–42.

Bicchieri, C. (2006). *The Grammar of Society: The Nature and Dynamics of Social Norms*. Cambridge, UK: Cambridge University Press.

Bies, R.J., and Moag, J.S. (1986). Interactional justice: Communication criteria of fairness. In R.J. Lewicki, B.H. Sheppard, and M.H. Bazerman (eds), *Research on Negotiation in Organizations* (Vol. 1, pp. 43–55). Greenwich, CT: JAI Press.

Blakemore, S.J. (2004). Social cognitive neuroscience: Where are we heading? *Trends in Cognitive Sciences*, *8*(5), 216–222.

Blanton, H., and Jaccard, J. (2008). Unconscious racism: A concept in pursuit of a measure. *Annual Review of Sociology*, *34*, 277–297.

Blount, S. (1995). When social outcomes aren't fair: The effect of causal attributions on preferences. *Organizational Behavior and Human Decision Processes*, *63*(2), 131–144.

Boden, M.A. (1998). Creativity and artificial intelligence. *Artificial Intelligence*, *103*(1/2), 347–356.

Bohnet, I., and Zeckhauser, R. (2004). Social comparisons in ultimatum bargaining. *Scandinavian Journal of Economy*, *106*(3), 495–510.

Bolino, M.C., and Grant, M.A. (2016). The bright side of being prosocial at work, and the dark side, too: A review and agenda for research on other-oriented motives, behavior, and impact in organizations. *Academy of Management Annals*, *10*(1), 599–670.

Boorman, E.D., Behrens, T.E.J., Woolrich, M.W., and Rushworth, M.F.S. (2009). How green is the grass on the other side? Frontopolar cortex and the evidence in favor of alternative courses of action. *Neuron*, *62*(5), 733–743.

Borg, J.S., Hynes, C., and Van Horn, J. (2006). Consequences, action, and intention as factors in moral judgments: An fMRI investigation. *Journal of Cognitive Neuroscience*, *18*(5), 803–817.

Borg, J.S., Lieberman, D., and Kiehl, K.A. (2008). Infection, incest, and iniquity: Investigating the neural correlates of disgust and morality. *Journal of Cognitive Neuroscience*, *20*(9), 1529–1546.

Borg, J.S., Sinnott-Armstrong, W., Calhoun, V.D., and Kiehl, K.A. (2011). Neural basis of moral verdict and moral deliberation. *Social Neuroscience*, *6*(4), 398–413.

Botvinick, M., and Braver, T. (2015). Motivation and cognitive control: From behavior to neural mechanism. *Annual Review of Psychology, 66,* 83–113.

Bowie, N.E. (1999). *Business Ethics: A Kantian Perspective.* Malden, MA: Basil Blackwell Publishers.

Boyatzis, R., and McKee, A. (2005). *Resonant Leadership.* Boston, MA: Harvard Business School Press.

Boyatzis, R.E., Passarelli, A.M., and Koenig, K. et al. (2012). Examination of the neural substrates activated in memories of experiences with resonant and dissonant leaders. *The Leadership Quarterly, 23*(2), 259–272.

Braeutigam, S. (2005). Neuroeconomics: From neural systems to economic behavior. *Brain Research Bulletin, 67*(5), 355–360.

Braeutigam, S. (2014). Organizational neuroscience: A new frontier for magnetoencephalography. In S. Supek and C.J. Aine (eds), *Magnetoencephalography. From Signals to Dynamic Cortical Networks* (pp. 743–748). Berlin: Springer.

Breitner, H.C., Aharon, I., and Kahneman, D. et al. (2001). Functional imaging of neural responses to expectancy and experience of monetary gains and losses. *Neuron, 30*(2), 619–639.

Bromberg-Martin, E.S., Matsumoto, M., and Hikosaka, O. (2010). Dopamine in motivational control: Rewarding, aversive, and alerting. *Neuron, 68*(5), 815–834.

Bromiley, O., and Cummings, L.L. (1996). Transaction costs in organizations with trust. In R. Bies, R. Lewicki, and B. Sheppard (eds), *Research on Negotiation in Organizations* (Vol. 5, pp. 219–247). Greenwich, CT: JAI Press.

Buckholtz, J.W., and Marois, R. (2012). The roots of modern justice: Cognitive and neural foundations of social norms and their enforcement. *Nature Neuroscience, 5*(5), 655–661.

Buckner, R.L., Andrews-Hanna, J.R., and Schacter, D.L. (2008). The brain default network: Anatomy, function, and relevance to disease. *Annals of the New York Academy of Sciences, 1124,* 1–38.

Buhle, J.T., Silvers, J.A., and Wager, T.D. et al. (2014). Cognitive reappraisal of emotion: A meta-analysis of human neuroimaging studies. *Cerebral Cortex, 24*(11), 2981–2990.

Bunge, S.A., and Wendelken, C. (2009). Comparing the bird in the hand with the ones in the bush. *Neuron, 62*(5), 609–611.

Burgelman, R.A. (1983). Corporate entrepreneurship and strategic management: Insights from a process study. *Management Science, 29*(12), 1349–1364.

Bush, G., Vogt, B.A., and Holmes, J. et al. (2002). Dorsal anterior cingulate cortex: A role in reward-based decision making. *PNAS*, *99*(1), 523–528.

Butler, M.J.R. (2014). Operationalizing interdisciplinary research: A model of co-production in organizational neuroscience. *Frontiers in Human Neuroscience*, *7*(720), 1–3.

Butler, M.J.R., and Senior, C. (2007). Toward an organizational cognitive neuroscience. *Annals of the New York Academy of Sciences*, *1118*, 1–17.

Butler, M.J.R., O'Broin, H.L.R., Lee, N., and Senior, C. (2016). How organizational cognitive neuroscience can deepen understanding of managerial decision-making: A review of the recent literature and future directions. *International Journal of Management Reviews*, *18*(4), 542–559.

Butterfield, K.D., Trevino, L.K., and Weaver, G.R. (2000). Moral awareness in business organizations: Influences of issue-related and social context factors. *Human Relations*, *53*(7), 981–1018.

Button, K.S., Ioannidis, J.P.A., and Mokrysz, C. et al. (2013). Power failure: Why small sample size undermines the reliability of neuroscience. *Nature Reviews Neuroscience*, *14*(5), 365–376.

Cabeza, R., and Nyberg, L. (1997). Imaging cognition: An empirical review of PET studies with normal subjects. *Journal of Cognitive Neuroscience*, *9*(1), 1–26.

Caceda, R., James, G.A., Gutman, D.A., and Kilts, C.D. (2015). Organization of intrinsic functional brain connectivity predicts decisions to reciprocate social behavior. *Behavioral Brain Research*, *292*, 478–483.

Cacioppo, J.T. (2002). Social neuroscience: Understanding the pieces fosters understanding the whole and vice versa. *American Psychologist*, *57*(11), 819–831.

Cacioppo, J.T., Amaral, D.G., and Blanchard, J.J. et al. (2007). Social neuroscience: Progress and implications for mental health. *Perspectives on Psychological Science*, *2*(2), 99–123.

Caldu, X., and Dreher J.C. (2007). Hormonal and genetic influences on processing reward and social information. *Annals of the New York Academy of Sciences*, *1118*, 43–73.

Camerer, C.F. (1999). Behavioral economics: Reunifying psychology and economics. *Proceedings of the National Academy of Sciences in the United States of America*, *96*(19), 10575–10577.

Camerer, C.F. (2003). *Behavioral Game Theory: Experiments in Strategic Interaction*. Princeton, NJ: Princeton University Press.

Camerer, C.F. (2007). Neuroeconomics: Using neuroscience to make economic predictions. *The Economic Journal*, *117*(519), C26–C42.

Camerer, C.F., Bhatt, M., and Hsu, M. (2007). Neuroeconomics: Illustrated by the study of ambiguity aversion. In B.B. Frey and A. Stutzer (eds), *Economics and Psychology: A Promising New Cross-disciplinary Field* (pp. 113–151). Cambridge, MA: MIT Press.

Camerer, C.F., and Loewenstein, G. (2004). Behavioral economists: Past, present, future. In C. Camerer, G. Loewenstein, and M. Rabin (eds), *Advances in Behavioral Economics* (pp. 3–51). Princeton, NJ: Princeton University Press.

Camerer, C.F., Loewenstein, G., and Prelec, D. (2004). Neuroeconomics: Why economics needs brains. *The Scandinavian Journal of Economics, 106*(3), 555–579.

Camerer, C.F., Loewenstein, G., and Prelec, D. (2005). Neuroeconomics: How neuroscience can inform economics. *Journal of Economic Literature, 43*(1), 9–64.

Camerer, C.F., and Mobbs, D. (2017). Differences in behavior and brain activity during hypothetical and real choices. *Trends in Cognitive Sciences, 21*(1), 46–56.

Camerer, C.F., and Weber, M. (1992). Recent developments in modeling preferences: Uncertainty and ambiguity. *Journal of Risk and Uncertainty, 5*(4), 325–370.

Cameron, D.C., Paynec, B.K., and Sinnott-Armstrong, W. et al. (2017). Implicit moral evaluations: A multinomial modeling approach. *Cognition, 158*(January), 224–241.

Campanha, C., Minati, L., Fregni, F., and Boggio, P.S. (2011). Responding to unfair offers made by a friend: Neurological activity changes in the anterior medial prefrontal cortex. *Journal of Neuroscience, 31*(43), 15569–15574.

Cappelen, A.W., Eichele, T., and Hugdahl, K. et al. (2014). Equity theory and fair inequality: A neuroeconomic study. *Proceedings of the National Academy of Sciences, 111*(43), 15368–15372.

Carter, C.S. (2014). Oxytocin pathways and the evolution of human behavior. *Annual Review of Psychology, 65*, 17–39.

Casebeer, W.D. (2003). Moral cognition and its neural constituents. *Nature Reviews Neuroscience, 4*, 841–846.

Caselli, R.J. (2002). Creativity: An organizational schema. *Cognitive and Behavioral Neurology, 22*(3), 143–154.

Cerasoli, C.P., Nicklin, J.M., and Ford, M.T. (2014). Intrinsic motivation and extrinsic incentives jointly predict performance: A 40-year meta-analysis. *Psychological Bulletin, 140*(4), 980–1008.

Chakravarty, A. (2010). The creative brain: Revisiting concepts. *Medical Hypotheses, 74*(3), 606–612.

Chakroff, C., Russell, P.S., Piazza, J., and Young, L. (2016). From impure to harmful: Asymmetric expectations about immoral agents. *Journal of Experimental Social Psychology*, *69*(March), 201–209.

Chamorro-Premuzic, T. (2013). Does money really affect motivation? A review of the literature. *Harvard Business Review*, April 10, 7–15.

Chang, L.J., and Sanfey, A.G. (2013). Great expectations: Neural computations underlying the use of social norms in decision-making. *Social Cognitive and Affective Neuroscience*, *8*(3), 277–284.

Chaturvedi, S., Arvey, R.D., Zhang, Z., and Christoforou, P.T. (2011). Genetic underpinnings of transformational leadership: The mediating role of dispositional hope. *Journal of Leadership and Organizational Studies*, *18*(4), 469–479.

Chaturvedi, S., Zyphur, M.J., and Arvey, R.D. et al. (2012). The heritability of emergent leadership: Age and gender as moderating factors. *The Leadership Quarterly*, *23*(2), 219–232.

Chaudhuri, A. (2011). Sustaining cooperation in laboratory public goods experiments: A selective survey of the literature. *Experimental Economics*, *14*(1), 47–83.

Chen, F.S., Kumsta, R., and Heinrichs, M. (2011). Oxytocin-induced goodwill is not a fixed pie. *Proceedings of the National Academy of Sciences*, *108*(13), E45.

Chermahini, S.A., and Hommel, B. (2010). The (b)link between creativity and dopamine: Spontaneous eye blink rates predict and dissociate divergent and convergent thinking. *Cognition*, *115*(3), 458–465.

Chi, R.P., and Snyder, A.W. (2012). Brain stimulation enables the solution of an inherently difficult problem. *Neuroscience Letters*, *515*(2), 121–124.

Christensen, L.J., Peirce, E., and Hartman, L.P. et al. (2007). Ethics, CSR, and sustainability education in the Financial Times top 50 business schools: Baseline data and future research direction. *Journal of Business Ethics*, *73*(4), 347–368.

Cocchi, L., Zaleski, A., Fornito, A., and Mattingley, J.B. (2013). Dynamic cooperation and competition between brain systems during cognitive control. *Trends in Cognitive Sciences*, *17*(10), 494–501.

Cohen, D. (1968). Electroencephalography: Evidence of magnetic fields produced by alpha-rhythms currents. *Science*, *161*(3841), 784–786.

Cohen, D. (1972). Magnetoencephalography: Detection of the brain's electrical activity with a superconducting magnetometer. *Science*, *175*(4023), 664–666.

Cohen, J.D. (2005). The vulcanization of the human brain: A neural perspective on interactions between cognition and emotion. *Journal of Economic Perspectives*, *19*(4), 3–24.

Cohen, J.D., McClure, S.M., and Yu, A. (2007). Should I stay or should I go? How the human brain manages the trade-off between exploitation and exploration. *Philosophical Transactions of the Royal Society*, *362*(1481), 933–942.

Colarelli, S.M., and Arvey, R.D. (2015). *The Biological Foundations of Organizational Behavior*, Chicago, IL: University of Chicago Press.

Cole, M.W., and Schneider, W. (2007). The cognitive control network: Integrated cortical regions with dissociable functions. *NeuroImage*, *37*(1), 343–360.

Colquitt, J.A., Conlon, D.E., and Wesson, M.J. (2001). Justice at the millennium: A meta-analytic review of 25 years of organizational justice research. *Journal of Applied Psychology*, *86*(3), 425–445.

Conger, J.A., and Kanungo, R.N. (1987). Toward a behavioral theory of charismatic leadership in organizational settings. *Academy of Management Review*, *12*(4), 637–647.

Conger, J.A., and Kanungo, R.N. (1988). *Charismatic Leadership: The Elusive Factor in Organizational Effectiveness*. San Francisco, CA: Jossey-Bass.

Contreras-Huerta, L.S., Baker, K.S., and Reynolds, K.J. (2013). Racial bias in neural empathic responses to pain. *PLOS One*, *8*(12), 1–10.

Cornelissen, P.L., Kringelbach, M.L., and Ellis, A. et al. (2009). Activation of the left inferior frontal gyrus in the first 200 ms of reading: Evidence from magnetoencephalography (MEG). *PLOS One*, *4*(4), 1–13.

Corradi-Dell'Acqua, C., Civai, C., Rumiati, R.I., and Fink, G.R. (2012). Disentangling self and fairness-related neural mechanisms involved in the ultimatum game: An fMRI study. *Social Cognitive and Affective Neuroscience*, *8*(4), 424–431.

Cristoff, K., Gordon, A.M., and Smallwood, J. et al. (2009). Experience sampling during fMRI reveals default network and executive system contributions to mind wandering. *Proceedings of the National Academy of Sciences*, *106*(21), 8719–8724.

Crocker, J., Canevello, A., and Brown, A.A. (2017). Social motivation: Costs and benefits of selfishness and otherishness. *Annual Review of Psychology*, *68*, 299–325.

Crockett, M.J. (2009). The neurochemistry of fairness: Clarifying the link between serotonin and prosocial behavior. *Annals of the New York Academy of Sciences*, *1167*, 76–86.

Crockett, M.J., Clark, L., Hauser, M.D., and Robbins, T.W. (2010). Serotonin selectively influences moral judgment and behavior through effects on harm aversion. *Proceedings of the National Academy of Sciences*, *107*(40), 17433–17438.

Crockett, M.J., Clark, L., and Lieberman, M.D. et al. (2010). Impulsive choice and altruistic punishment are correlated and increase in tandem with serotonin depletion. *Emotion, 10*(6), 855–862.

Crockett, M.J., Clark, L., and Tabibnia et al. (2008). Serotonin modulates behavioral reactions to unfairness. *Science, 320*(5884), 1739–1743.

Cropanzano, R., and Becker, W.J. (2013). The promise and peril of organizational neuroscience: Today and tomorrow. *Journal of Management Inquiry, 22*(3), 306–310.

Cropanzano, R., and Wright, T.A. (2011). The impact of organizational justice on occupational health. In J.C. Quick and L.E. Tetrick (eds), *Handbook of Occupational Health Psychology* (pp. 205–219). Washington, DC: American Psychological Association.

Cropanzano, R.S., Massaro, S., and Becker, W.J. (2017). Deontic justice and organizational neuroscience. *Journal of Business Ethics, 144*(4), 733–754.

Cunningham, W.A., Johnson, M.K., and Raye, C.L. et al. (2004). Separable neural components in the processing of black and white faces. *Psychological Science, 15*(12), 806–813.

Cunningham, W.A., Zelazo, P.D., Packer, D.J., and Van Bavel, J.J. (2007). The iterative reprocessing model: A multilevel framework for attitudes and evaluation. *Social Cognition, 25*(5), 736–760.

Cushman, F. (2013). Action, outcome, and value: A dual system framework for morality. *Personality and Social Psychology Review, 17*(3), 273–292.

Cushman, F., Murray, D., and Gordon-McKeon, S. et al. (2012). Judgment before principle: Engagement of the frontoparietal control network in condemning harms of omission. *Social Cognitive and Affective Neuroscience, 7*(8), 888–895.

Damasio, A.R. (1994). *Descartes' Error: Emotion, Reason and the Human Brain.* New York: Penguin.

Damasio, A.R., Everitt, B.J., and Bishop, D. (1996). The somatic marker hypothesis and the possible functions of the prefrontal cortex. *Philosophical Transactions: Biological Sciences, 351*(1346), 1413–1420.

Dasborough, M.T. (2006). Cognitive asymmetry in employee emotional reactions to leadership behaviors. *Leadership Quarterly, 17*(2), 163–178.

Da Silva, F.L. (2013). EEG and MEG: Relevance to neuroscience. *Neuron, 80*(5), 1112–1128.

Davidson, R.J., Jackson, D.C., and Kalin, N.H. (2000). Emotion, plasticity, context, and regulation: Perspectives from affective neuroscience. *Psychological Bulletin, 126*(6), 890–909.

Daw, N.D., and Shohami, D. (2008). The cognitive neural science of motivation and learning. *Social Cognition, 26*(5), 593–620.

Daw, N.D., O'Doherty, J.P., and Dayan, P. et al. (2006). Cortical substrates for exploratory decisions in humans. *Nature, 441*(7095), 876–879.

Dawes, R.M., and Messick, D.M. (2000). Social dilemmas. *International Journal of Psychology, 35*(2), 111–116.

Dawkins, R. (1976). *The Selfish Gene.* Oxford: Oxford University Press.

Deak, A. (2011). Brain and emotion: Cognitive neuroscience of emotions. *Review of Psychology, 18*(2), 71–80.

De Breu, C.K.W. (2012). Oxytocin modulates cooperation within and competition between groups: An integrative review and research agenda. *Hormones and Behavior, 61*(3), 419–428.

De Breu, C.K.W., and Cret, M.E. (2016). Oxytocin conditions intergroup relations through upregulated in-group empathy, cooperation, conformity, and defense. *Biological Psychiatry, 79*(3), 165–173.

De Breu, C.K.W., Greer, L.L., and Van Kleef, G.A. et al. (2011a). Oxytocin promotes human ethnocentrism. *Proceedings of the National Academy of Sciences, 108*(13), 1262–1266.

De Breu, C.K.W., Greer, L.L., and Van Kleef, G.A. et al. (2011b). Reply to Chen et al.: Perhaps goodwill is unlimited, but oxytocin-induced goodwill is not. *Proceedings of the National Academy of Sciences, 108*(13), E46.

Decety, J., and Cowell, J.M. (2014). The complex relation between morality and empathy. *Trends in Cognitive Sciences, 18*(7), 337–339.

Decety, J., and Jackson, P.L. (2004). The functional architecture of human empathy. *Behavioral and Cognitive Neuroscience Reviews, 3*(2), 71–100.

Decety, J., and Jackson, P.L. (2006). A social-neuroscience perspective on empathy. *Current Directions in Psychological Science, 15*(2), 54–58.

Decety, J., Jackson, P.L., and Sommerville, J.A. et al. (2004). The neural bases of cooperation and competition: An fMRI investigation. *NeuroImage, 23*(2), 744–751.

Deci, E.L. (1971). Effects of externally mediated rewards on intrinsic motivation. *Journal of Personality and Social Psychology, 18*(1), 105–115.

Deci, E., and Ryan, R. (1985). *Intrinsic Motivation and Self-Determination in Human Behavior.* New York: Plenum Press.

Deci, E.L., Koestner R., and Ryan R.M. (1999). A meta-analytic review of experiments examining the effects of extrinsic rewards on intrinsic motivation. *Psychological Bulletin, 125*(6), 627–668.

Declerck, C.H., Boone, C., and Emonds, G. (2013). When do people cooperate? The neuroeconomics of prosocial decision making. *Brain and Cognition*, *81*(1), 95–117.

De Holan, P.M. (2014). It's all in your head: Why we need neuro-entrepreneurship. *Journal of Management Inquiry*, *23*(1), 93–97.

Delgado, M.R., Frank, R.H., and Phelps, E.A. (2005). Perceptions of moral character modulate the neural systems of reward during the trust game. *Nature Neuroscience*, *6*(11), 1611–1618.

De Martino, B., Kumaran, D., Seymour, B., and Dolan, R.J. (2006). Frames, biases, and rational decision-making in the human brain. *Science*, *313*(5787), 684–687.

De Neve, J.E., Mikhaylov, S., and Dawes, C.T. et al. (2013). Born to lead? A twin design and genetic association study of leadership role occupancy. *The Leadership Quarterly*, *24*(1), 45–60.

De Quervain, D.J.F., Fischbacher, U., and Treyer, V. (2004). The neural basis of altruistic punishment. *Science*, *305*(5688), 1254–1258.

Deutsch, M. (1975). Equity, equality and need: What determines which value will be used as the basis of distributive justice? *Journal of Social Issues*, *31*(3), 137–149.

Deutsch, M. (1985). *Distributive Justice: Social Psychological Perspective*. New Haven, CT: Yale University Press.

De Vignemont, F., and Singer, T. (2006). The empathic brain: How, when and why? *Trends in Cognitive Sciences*, *10*(10), 435–441.

De Waal, F. (1996). *Good Natured: The Origins of Right and Wrong in Humans and Other Animals*. Cambridge, MA: Harvard University Press.

Dickhaut, J., McCabe, K., and Nagode, J.C. et al. (2003). The impact of the certainty context on the process of choice. *Proceedings of the National Academy of Sciences*, *100*(6), 3536–3541.

Di Domenico, S.I., and Ryan, R.M. (2017). The emerging neuroscience of intrinsic motivation: A new frontier in self-determination research. *Frontiers in Human Neuroscience*, *11*(March), 145–160.

Dietrich, A. (2004a). The cognitive neuroscience of creativity. *Psychological Bulletin and Review*, *11*(6), 1011–1026.

Dietrich, A. (2004b). The neurocognitive mechanisms underlying the experience of flow. *Conscientiousness and Cognition*, *13*(4), 746–761.

Dietrich, A., and Kanso, B. (2010). A review of EEG, ERP, and neuroimaging studies of creativity and insight. *Psychological Bulletin*, *136*(5), 822–848.

Dimoka, A. (2010). What does the brain tell us about trust and distrust? Evidence from a functional magnetic neuroimaging study. *MIS Quarterly*, *24*(2), 373–396.

Dulebohn, J.H., Conlon, D.E., and Sarinopoulos, I. et al. (2009). The biological bases of unfairness: Neuroimaging evidence for the distinctiveness of procedural and distributive justice. *Organizational Behavior and Human Decision Processes, 110*(2), 140–151.

Dulebohn, J.H., Davison, R.B., Lee, S.A., and Sarinopoulos, I. (2016). Gender differences in justice evaluations. *Journal of Applied Psychology, 101*(2), 151–170.

Dulleck, U., Ristl, A., Schaffner, M., and Torgler (2011). Heart rate variability, the autonomic nervous system, and neuroeconomic experiments. *Journal of Neuroscience, Psychology, and Economics, 4*(2), 117–124.

Dunne, D., and Martin, R. (2006). Design thinking and how it will change management education: An interview and discussion. *Academy of Management Learning and Education, 5*(4), 512–523.

Ellamil, M., Dobson, C., Beeman, M., and Christoff, K. (2012). Evaluative and generative modes of thought during the creative process. *NeuroImage, 59*(2), 1783–1794.

Ernst, M., and Paulus, M.P. (2005). Neurobiology of decision making: A selective review from a neurocognitive and clinical perspective. *Biological Psychiatry, 58*(8), 597–604.

Evans, J. St. B.T. (2008). Dual-process accounts of reasoning, judgment and social cognition. *Annual Review of Psychology, 59*, 255–278.

Evans, J. St. B.T. (2009). How many dual-process theories do we need? One, two, or many? In J. St. B.T. Evans and K. Frankish (eds), *In Two Minds: Dual Processes and Beyond* (pp. 33–35). New York: Oxford University Press.

Fan, Y., Duncan, N.W., De Greck, M., and Northoff, G. (2011). Is there a core neural network in empathy? An fMRI based quantitative meta-analysis. *Neuroscience and Biobehavioral Reviews, 35*(3), 903–911.

Farah, M.J. (2005). Neuroethics: The practical and the philosophical. *Trends in Cognitive Sciences, 9*(1), 34–40.

Farah, M.J. (2012). Neuroethics: The ethical, legal, and societal impact of neuroscience. *Annual Review of Psychology, 63*, 571–591.

Fehr, E., and Camerer, C.F. (2007). Social neuroeconomics: The neural circuitry of social preferences. *Trends in Cognitive Sciences, 11*(10), 419–427.

Fehr, E., and Fischbacher, U. (2004). Social norms and human cooperation. *Trends in Cognitive Sciences, 8*(4), 185–190.

Fehr, E., and Gächter, S. (2000). Fairness and retaliation: The economics of reciprocity. *Journal of Economic Perspectives, 14*(3), 159–181.

Fehr, E., and Gächter, S. (2002). Altruistic punishment in humans. *Nature, 415*(7021), 137–140.

Fehr, E. and Schmidt, K.M. (1999). A theory of fairness, competition, and cooperation. *The Quarterly Journal of Economics*, *114*(3), 817–868.

Fehr, E., Fischbacher, U., and Kosfeld, M. (2005). Neuroeconomic foundations of trust and social preferences: Initial evidence. *American Economic Review*, *95*(2), 346–351.

Feng, C., Hackett, P.D., and DeMarco, A.C. et al. (2015). Oxytocin and vasopressin effects on the neural response to social cooperation are modulated by sex in humans. *Brain Imaging and Behavior*, *9*(4), 754–764.

Feng, C., Luo, Y.J., and Krueger, F. (2015). Neural signatures of fairness-related normative decision making in the ultimatum game: A coordinated meta-analysis. *Human Brain Mapping*, *36*(2), 591–602.

Ferrell, O.C., Fraedrich, J., and Ferrell, L. (2017). *Business Ethics: Ethical Decision Making and Cases*. Boston, MA: Cengage Learning.

Fetchenhauer, D., and Huang, X. (2004). Justice sensitivity and distributive decisions in experimental games. *Personality and Individual Differences*, *36*(5), 1015–1029.

Fett, A.K.J., Gromann, P.M., and Giampietro, V. et al. (2014). Default distrust? An fMRI investigation of the neural development of trust and cooperation. *Social Cognitive Affective Neuroscience*, *9*(4), 395–402.

Fiedler, F.E. (1964). A contingency model of leadership effectiveness. *Advances in Social Experimental Psychology*, *1*, 149–190.

Fiedler, F.E. (1967). *A Theory of Leadership Effectiveness*. New York: McGraw-Hill.

Fink, A., Grabner, R.H., and Gebauer, D. et al. (2010). Enhancing creativity by means of cognitive stimulation. Evidence from an fMRI study. *NeuroImage*, *52*(4), 1687–1695.

Fleishman, E.A. (1953a). The measurement of leadership attitudes in industry. *Journal of Applied Psychology*, *37*(3), 153–158.

Fleishman, E.A. (1953b). The description of supervisory behavior. *Journal of Applied Psychology*, *37*(1), 1–6.

Fleishman, E.A. (1957). A leader behavior description for industry. In R.M. Stogdill and A.E. Coons (eds), *Leader Behavior: Its Description and Measurement*, Columbus, OH: The Ohio State University, Bureau of Business Research.

Fliessbach, K., Phillips, C.B., and Trautner, P. et al. (2012). Neural responses to advantageous and disadvantageous inequity. *Frontiers in Human Neuroscience*, *6*(June), 1–9.

Folger, R.G. (1998). Fairness as moral virtue. In M. Schminke (ed.), *Managerial Ethics*: *Moral Management of People and Processes* (pp. 13–34). Mahwah, NJ: Lawrence Erlbaum Associates.

Folger, R.G. (2001). Fairness as deonance. In S. Gilliland, D.D. Steiner, and D. Skarlicki (eds), *Theoretical and Cultural Perspectives on Organizational Justice* (pp. 3–33). Greenwich, CT: Information Age Publishing.

Folger, R.G., and Glerum, D.R. (2015). Justice and deonance: "You ought to be fair." In R. Cropanzano and M.A. Ambrose (eds), *Oxford Handbook of Justice in Work Organizations* (pp. 331–350). Oxford: Oxford University Press.

Forbes, C.A., and Grafman, J. (2010). The role of the human prefrontal cortex in social cognition and moral judgment. *Annual Review of Neuroscience, 33*, 299–324.

Fornasier, M., and Pitolli, F. (2008). Adaptive iterative thresholding algorithms for magnetoencephalography (MEG). *Journal of Computational and Applied Mathematics, 221*(2), 386–395.

Forstmann, B.U., Ratcliff, R., and Wagenmakers, E.J. (2016). Sequential sampling models of cognitive neuroscience: Advantages, applications, and extensions. *Annual Review of Psychology, 67*, 641–666.

Forsyth, D.R. (1980). A taxonomy of ethical ideologies. *Journal of Personality and Social Psychology, 39*(1), 175–184.

Forsyth, D.R. (1985). Individual differences in information integration during moral judgment. *Journal of Personality and Social Psychology, 49*(1), 264–272.

Frost, C.J., and Lumia, A.R. (2012). The ethics of neuroscience and the neuroscience of ethics: A phenomenological-existential approach. *Science and Engineering Ethics, 18*(3), 457–474.

Fukuyama, F. (1995). *Trust: The Social Virtues and the Creation of Prosperity*. New York: Free Press.

Funk, C.M., and Gazzaniga, M.S. (2009). The functional brain architecture of human morality. *Current Opinion in Neurobiology, 19*(6), 678–681.

Gainotti, G. (2012). Unconscious processing of emotions and the right hemisphere. *Neuropsychologia, 50*(2), 205–218.

Gallagher, H.L., and Frith, C.D. (2003). Functional imaging of theory of mind. *Trends in Cognitive Sciences, 7*(2), 77–83.

Gallese, V., Keysers, C., and Rizzolatti (2004). A unifying view of the basis of social cognition. *Trends in Cognitive Sciences, 8*(9), 396–403.

Gino, F. (2015). Understanding ordinary unethical behavior: Why people who value morality act immorally. *Current Opinion in Biological Sciences, 3*(June), 107–111.

Glenberg, A.M., Witt, J.K., and Metcalfe, J. (2013). From the revolution to embodiment: 25 years of cognitive psychology. *Perspectives on Psychological Science, 8*(5), 573–585.

Glimcher, P.W. (2003). The neurobiology of visual-saccadic decision making. *Annual Review* of *Neuroscience, 26*(1), 133–179.

Glimcher, P.W., Kable, J.W., and Louie, K. (2007). Neuroeconomics studies of impulsivity: Now or just as soon as possible? *American Economic Review, 97*(2), 142–147.

Golby, A.J., Gabrieli, J.D.E., Chian, J.Y., and Eberhardt, J.L. (2001). Differential fusiform responses to same and other race faces. *Nature Neuroscience, 4*(8), 845–850.

Gold, J.S., and Shadlen, M.N. (2007). The neural basis of decision making. *Annual Review of Psychology, 30*, 535–574.

Goldstein, W.M., and Weber, E.U. (1997). Content and discontent: Indications and implications of domain specificity in preferential decision making. In W.M. Goldstein, and R.M. Hogarth (eds), *Research on Judgment and Decision Making* (pp. 566–617). Cambridge, UK: Cambridge University Press.

Goleman, D. (2006). *Emotional Intelligence.* New York: Bantam Books.

Goleman, D., and Boyatzis, R. (2008). Social intelligence and the biology of leadership. *Harvard Business Review,* (September), 1–8.

Graen, G. (1976). Role-making processes within complex organizations. In M.D. Dunnette (ed.), *Handbook of Industrial and Organizational Psychology* (pp. 1201–1245). Chicago, IL: Rand McNally.

Grandey, A.A. (2000). Emotional regulation in the workplace: A new way to conceptualize emotional labor. *Journal of Occupational Health Psychology, 5*(1), 95–110.

Grandey, A.A., and Gabriel, A.S. (2015). Emotional labor at a crossroads: Where do we go from here? *Annual Review of Organizational Psychology and Organizational Behavior, 2*, 323–349.

Granovetter, M. (1985). Economic action and social structure: The problem of embeddedness. *American Journal of Sociology, 91*(3), 481–510.

Grant, A.M. (2008). Does intrinsic motivation fuel the prosocial fire? Motivational synergy in predicting persistence, performance, and productivity. *Journal of Applied Psychology, 93*(1), 48–58.

Grasso, M. (2013). Climate ethics: With a little help from moral cognitive neuroscience. *Environmental Politics, 22*(3), 377–393.

Greely, H. (2007). On neuroethics. *Science, 318*(5850), 533.

Greenberg, J. (1987). A taxonomy of organizational justice theories. *Academy of Management Review, 12*(1), 9–22.

Greenberg, J. (1990). Organizational justice: Yesterday, today, and tomorrow. *Journal of Management, 16*(2), 399–432.

Greene, J.D. (2003). From neural "is" to moral "ought": What are the moral implications of neuroscientific moral psychology? *Nature Reviews Neuroscience, 4*(10), 846–850.

Greene, J.D. (2008). The secret joke of Kant's soul. In S. Sinnott-Armstrong (ed.), *The Neuroscience of Morality: Emotion, Brain Disorders, and Development* (pp. 35–80). Cambridge, MA: MIT Press.

Greene, J.D. (2009). Dual-process morality and the personal/impersonal distinction: A reply to McGuire, Langdon, Coltheart, and Mackenzie. *Journal of Experimental Social Psychology*, *45*(3), 581–584.

Greene, J.D. (2014). Beyond point-and-shoot morality: Why cognitive (neuro) science matters for ethics. *Ethics*, *124*(4), 695–726.

Greene, J.D. (2015). The rise of moral cognition. *Cognition*, *135*(February), 39–42.

Greene, J.D., Nystrom, L.E., and Engell, A.D. et al. (2004). The neural bases of cognitive conflict and control in moral judgment. *Neuron*, *44*(2), 389–400.

Greene, J.D., Sommerville, R.B., and Nystrom, L.E. et al. (2001). An fMRI investigation of emotional engagement in moral judgment. *Science*, *293*(5537), 2105–2108.

Guilford, J.P. (1967). *The Nature of Human Intelligence*. New York: McGraw-Hill.

Güroğlu, B., Van den Bos, W., and Crone, E.A. (2009). Fairness considerations: Increasing understanding of intentionality in adolescence. *Journal of Experimental Child Psychology*, *104*(4), 398–409.

Güroğlu, B., Van den Bos, W., and Van Dijk, E. et al. (2011). Dissociable brain networks involved in development of fairness considerations: Understanding intentionality behind unfairness. *NeuroImage*, *57*(2), 634–641.

Güth, W., Schmittberger, R., and Schwarze, B. (1982). An experimental analysis of ultimatum bargaining. *Journal of Economic Behavior and Organization*, *3*(4), 367–388.

Gutsell, J.N., and Inzlicht, M. (2010). Empathy constrained: Prejudice predicts reduced mental simulation of actions during observation of outgroups. *Journal of Experimental Social Psychology*, *46*(5), 841–845.

Haas, B.W., Ishak, A., Anderson, I.W., and Fikowski, M.M. (2015). The tendency to trust is reflected in human brain structure. *NeuroImage*, *107*(February), 175–181.

Hahn, T., Notebaert, K. and Ander, C. et al. (2015). How to trust a perfect stranger: Predicting initial trust behavior from resting-state brain-electrical connectivity. *Social Cognitive Affective Neuroscience*, *10*(6), 809–813.

Haidt, J. (2001). The emotional dog and its rational tail: A social intuitionist approach to moral judgment. *Psychological Review*, *108*(4), 814–834.

Haidt, J. (2003). The moral emotions. In R.J. Davidson, K.R. Scherer, and H.H. Goldsmith (eds), *Handbook of Affective Sciences* (pp. 852–870). Oxford: Oxford University Press.

Haidt, J. (2007). The new synthesis in moral psychology. *Science*, *316*(5827), 998–1002.

Haidt, J. (2008). Morality. *Perspectives on Psychological Science*, *3*(1), 65–72.

Haidt, J. (2012). *The Righteous Mind: Why Good People Are Divided by Politics and Religion*, New York: Pantheon.

Handy, T.C. (2005). *Event-related Potentials: A Methods Handbook*. Cambridge, MA: MIT Press.

Hannah, S.T., and Waldman, D.A. (2015). Neuroscience of moral cognition and conation in organizations. In D.A. Waldman and P.A. Balthazard (eds), *Organizational Neuroscience (Monographs in Leadership and Management*, (Vol. 7, pp. 233–255). Bingley, UK: Emerald Group Publishing Limited.

Hannah, S.T., Avolio, B.J., and May, D.R. (2011). Moral maturation and moral conation: A capacity approach to explaining moral thought and action. *Academy of Management Review*, *36*(4), 663–685.

Hannah, S.T., Balthazard, P.A., Waldman, D.A., and Jennings, P.L. (2013). The psychological and neural bases of leader self-complexity and effects on adaptive decision-making. *Journal of Applied Psychology*, *98*(3), 393–411.

Harle, K.M., Chang, L.J., Wout, M., and Sanfey, A.G. (2012). The neural mechanisms of affect infusion in social economic decision-making: A mediating role of the anterior insula. *NeuroImage*, *61*(1), 32–40.

Harlow, H.F. (1950). Learning and satiation of response in intrinsically motivated complex puzzle performance by monkeys. *Journal of Comparative and Physiological Psychology*, *43*(4), 289–294.

Harris, L.T., and Fiske, S.F. (2006). Dehumanizing the lowest of the low: Neuroimaging responses to extreme out-groups. *Psychological Science*, *17*(10), 847–853.

Hart, A.J., Whalen, P.J., and Shin, L.M. (2000). Differential response in the human amygdala to racial outgroup vs ingroup face stimuli. *NeuroReport*, *11*(11), 2351–2355.

Haruno, M., and Frith, C.D. (2010). Activity in the amygdala elicited by unfair divisions predicts social value orientation. *Nature Neuroscience*, *13*(2), 160–161.

Haruno, M., and Kawato, M. (2006). Different neural correlates of reward expectation and reward expectation error in the putamen and caudate nucleus during stimulus–action–reward association learning. *Journal of Neurophysiology*, *95*(2), 948–959.

Hatfield, E., Cacioppo, J.T., and Rapson, R.L. (1994). *Emotional Contagion*. New York: Cambridge University Press.

Healy, M.P., and Hodgkinson, G.P. (2014). Rethinking the philosophical and theoretical foundations of organizational neuroscience: A critical realist alternative. *Human Relations, 67*(7), 765–792.

Heaphy, E.D., and Dutton, J.E. (2008). Positive social interactions and the human body at work: Linking organizations and physiology. *Academy of Management Review, 33*(1), 137–162.

Hebb, D.O. (1949). *The Organization of Behavior: A Neuropsychological Theory*. New York: John Wiley and Sons.

Hellige, J.B. (1990). Hemispheric asymmetry. *Annual Review of Psychology, 41*, 55–80.

Hennessey, B.A., and Amabile, T.M. (2010). Creativity. *Annual Review of Psychology, 61*, 569–598.

Henson, R. (2006). Forward inference using functional neuroimaging: Dissociations versus associations. *Trends in Cognitive Sciences, 10*(2), 64–69.

Hersey, P., and Blanchard, K.H. (1969). Life cycle theory of leadership. *Training & Development Journal, 23*(5), 26–34.

Hersey, P., and Blanchard, K.H. (1977). *Management of Organizational Behavior: Utilizing Human Resources*. Third edition. Englewood Cliffs, NJ: Prentice Hall.

Hersey, P., and Blanchard, K.H. (1982). *Management of Organizational Behavior: Utilizing Human Resources*. Fourth edition. Englewood Cliffs, NJ: Prentice Hall.

Hewig, J., Kretschmer, N., and Trippe, R.H. et al. (2011). Why humans deviate from rational choice. *Psychophysiology, 48*(4), 507–514.

Hitlin, S., and Vaisey, S. (2013). The new sociology of morality. *Annual Review of Sociology, 39*, 51–68.

Hochschild, A.R. (1983). *The Managed Heart: Commercialization of Human Feeling*. Berkeley, CA: University of California Press.

Holroyd, C.B., and Yeung, N. (2012). Motivation of extended behaviors by anterior cingulate cortex. *Trends in Cognitive Sciences, 16*(2), 122–128.

Hölzel, B.K., Carmody, J., and Vangel, M. et al. (2011). Mindfulness practice leads to increases in regional brain gray matter density. *Psychiatry Research: Neuroimaging, 191*(1), 36–43.

Horvath, J., Forte, J., and Carter, O. (2015). Evidence that transcranial direct current stimulation (tDCS) generates little-to-no reliable neurophysiologic effect beyond MEP amplitude modulation in healthy human subjects: A systematic review. *Neuropsychologia, 66*(January), 213–236.

House, R.H. (1971). A path-goal theory of leader effectiveness. *Administrative Science Quarterly, 16*(3), 321–339.

Hsu, M., Anen, C., and Quartz, S.R. (2008). The right and the good: Distributive justice and neural encoding of equity and efficiency. *Science, 320*(5879), 1092–1095.

Hsu, M., Bhatt, M., and Adolphs, R. et al. (2005). Neural systems responding to degrees of uncertainty in human decision-making. *Science, 10*(5754), 1679–1683.

Huettel, S.A. (2006). Behavioral, but not reward, risk modulates activation of prefrontal, parietal, and insular cortices. *Cognitive, Affective, and Behavioral Neuroscience, 6*(2), 141–151.

Huettel, S.A., Stowe, C.J., and Gordon, E.M. et al. (2006). Neural signatures of economic preferences for risk and ambiguity. *Neuron, 49*(5), 765–775.

Iacoboni, M. (2009). Imitation, empathy, and mirror neurons. *Annual Review of Psychology, 60*, 653–670.

Ilardi, S.S., and Feldman, D. (2001). The cognitive neuroscience paradigm: A unifying metatheoretical framework for the science and practice of clinical psychology. *Journal of Clinical Psychology, 57*(9), 1067–1088.

Illes, J., and Bird, S.J. (2006). Neuroethics: A modern context for ethics in neuroscience. *Trends in Neuroscience, 29*(9), 511–517.

Innocenti, A., and Sirigu, A. (2012). *Neuroscience and the Economics of Decision Making.* New York: Routledge.

Ito, T.A., and Urland, G.R. (2003). Race and gender on the brain: Electrocortical measures of attention to the race and gender of multiply categorizable individuals. *Journal of Personality and Social Psychology, 85*(4), 616–626.

Ives-Deliperi, V.L., Solms, M., and Meintjes, E.M (2011). The neural substrates of mindfulness: An fMRI investigation. *Journal of Social Neuroscience, 6*(3), 231–242.

Izuma, K., Saito, D., and Sadato, N. (2008). Processing of social and monetary rewards in the human striatum. *Neuron, 58*(2), 284–294.

Jack, A.I., Boyatzis, R.E., and Khawaja, M.S. et al. (2013). Visioning in the brain: An fMRI study of inspirational coaching and mentoring. *Social Neuroscience, 8*(4), 369–384.

Jack, A.I., Dawson, A.J., and Norr, M.E. (2013). Seeing human: Distinct and overlapping neural signatures associated with two forms of dehumanization. *NeuroImage, 79*, 313–328.

Jack, A.I., Rochford, K.C., and Friedman, J.P. et al. (2017). Pitfalls in organizational neuroscience: A critical review and suggestions for future research. *Organizational Research Methods.* DOI: 10.1177/1094428117708857.

Jarosz, A.F., Colflesh, G.J., and Wiley, J. (2012). Uncorking the muse: Alcohol intoxication facilitates creative problem solving. *Conscious-ness and Cognition, 21*(1), 487–493.

Jausovec, N. (2000). Differences in cognitive processes between gifted, intelligent, creative, and average individuals while solving complex problems: An EEG study. *Intelligence, 28*(3), 213–237.

Jenkins, I.H., Brooks, D.J., and Nixon, P.D. et al. (1994). Motor sequence learning: A study with positron emission tomography. *Journal of Neuroscience, 14*(6), 3775–3790.

Jiang, J., Chen, C., and Dai, B. (2015). Leader emergence through interpersonal neural synchronization. *Proceedings of the National Academy of Sciences, 112*(14), 4274–4279.

Jin, J., Yu, L., and Ma, Q. (2015). Neural basis of intrinsic motivation: Evidence from event-related potentials. *Computational Intelligence and Neuroscience, 2015*(7), 1–6.

Johnson, A.M., Vernon, P.A., and McCarthy, J.M. et al. (1998). Nature vs. nurture: Are leaders born or made? A behavior genetic investigation of leadership style. *Twin Research, 1*(4), 216–223.

Johnson, N.D., and Mislin, A.A. (2011). Trust games: A meta-analysis. *Journal of Economic Psychology, 32*(5), 865–889.

Jones, T.M. (1991). Ethical decision making by individuals in organ-izations: An issue-contingent model. *Academy of Management Review, 16*(2), 366–395.

Jones, T.M., and Ryan, L.V. (1997). The link between ethical judgment and action in organizations: A moral approbation approach. *Organ-ization Science, 8*(6), 663–680.

Jones, T.M., and Ryan, L.V. (1998). The effect of organizational forces on individual morality: Judgment, moral approbation, and behavior. *Business Ethics Quarterly, 8*(3), 431–445.

Jung, R.E., Mead, B.S., Carrasco, J., and Flores, R.A. (2013). The structure of creative cognition in the human brain. *Frontiers in Human Neuroscience, 7*(July), 1–13.

Kable, W.J. (2011). The cognitive neuroscience toolkit for the neuro-economist: A functional overview. *Journal of Neuroscience, Psychol-ogy, and Economics, 4*(2), 63–84.

Kahneman, D. (2003). Mapping bounded rationality: Psychology for behavioral economics. *American Economic Review, 93*(5), 1449–1475.

Kahneman, D. (2011). *Thinking, Fast and Slow.* New York: Farrar, Straus and Giroux.

Kahneman, D., and Tversky, A. (1979). Prospect theory: An analysis of decision under risk. *Econometrica, 47*(2), 263–292.

Kahneman, D., and Tversky, A. (1984). Choices, values, and frames. *American Psychologist, 39*(4), 341–350.

Kahneman, D., Knetsch, J.L., and Thaler, R.H. (1990). Experimental tests of the endowment effect and the Coase theorem. *Journal of Political Economy*, *98*(6), 1325–1348.

Kahneman, D., Knetsch, J.L., and Thaler, R.H. (1991). Anomalies: The endowment effect, loss aversion, and status quo bias. *Journal of Economic Perspectives*, *5*(1), 193–206.

Kanat, M., Heinrichs, M., and Domes, G. (2014). Oxytocin and the social brain: Neural mechanisms and perspectives in human research. *Brain Research*, *1580*(September), 160–171.

Kaplan, A. (1964). *The Conduct of Inquiry: Methodology for Behavioral Science*, Scranton, PA: Chandler Publishing.

Kaplan, C.A., and Simon, H.A. (1990). In search of insight. *Cognitive Psychology*, *22*(3), 374–419.

Karni, E. and Safra, Z. (2002a). Individual sense of justice: A utility representation. *Econometrica*, *70*(1), 263–284.

Karni, E. and Safra, Z. (2002b). Intensity of the sense of fairness: Measurement and behavioral characterization. *Journal of Economic Theory*, *105*(2), 318–337.

Karni, E., Salmon, T., and Sopher, B. (2008). Individual sense of fairness: An experimental study. *Experimental Economics*, *11*(2), 174–189.

Kaufman, A.B., Kornilov, S.A., and Bristol, A.S. et al. (2010). The neurobiological foundations of creative cognition. In J.C. Kaufman and R.L. Sternberg (eds), *The Cambridge Handbook of Creativity* (pp. 216–232). New York: Cambridge University Press.

Kennerley, S.W., and Walton, M.E. (2011). Decision making and reward in frontal cortex: Complementary evidence from neuropsychological and neuropsychological studies. *Behavioral Neuroscience*, *125*(3), 297–317.

Kim, S.I. (2013). Neuroscientific model of motivational process. *Frontiers in Psychology*, *4*(Article 98), 1–12.

Kim, S.I., Reeve, J., and Bong, M. (2016). Introduction to motivational neuroscience. In S.I. Kim, J. Reeve, and M. Bong (eds), *Recent Developments in Neuroscience Research on Human Motivation. Advances in Motivation and Achievement* (Vol. 19, pp. 1–19). Bingley, UK: Emerald Group Publishing Limited.

King-Casas, B., Tomlin, D., and Anen, C. et al. (2005). Getting to know you: Reputation and trust in a two-person economic exchange. *Science*, *308*(5718), 78–83.

Klein, C. (2010). Philosophical issues in neuroimaging. *Philosophy Compass*, *5*(2), 186–198.

Klimoski, R., and Mohammed, S. (1994). Team mental model: Construct or metaphor? *Journal of Management*, *20*(2), 403–437.

Knight, F.H. (1921). *Risk, Uncertainty, and Profit*. New York: Houghton Mifflin.

Knoch, D., and Fehr, E. (2007). Resisting the power of temptations: The right prefrontal cortex and self-control. *Annals of the New York Academy of Sciences, 1104*, 123–134.

Knoch, D., Nitsche, M.A., and Fischbacher, U. et al. (2008). Studying the neurobiology of social interaction with transcranial direct current stimulation: The example of punishing unfairness. *Cerebral Cortex, 18*(9), 1987–1990.

Knoch, D., Pascual-Leone, A., and Myer, K. et al. (2006). Diminishing reciprocal fairness by disrupting the right prefrontal cortex. *Science, 314*(5800), 829–832.

Knutson, B., and Bossaerts, P. (2007). Neural antecedents of financial decisions. *Journal of Neuroscience, 27*(31), 8174–8177.

Kohlberg, L. (1969). Stage and sequence: The cognitive developmental approach to socialization. In D.A. Goslin (ed.), *Handbook of Socialization Theory and Research* (pp. 151–235). New York: Academic Press.

Komssi, S., and Kähkönen, S. (2006). The novelty value of the combined use of electroencephalography and transcranial magnetic stimulation for neuroscience research. *Brain Research Reviews, 52*(1), 183–192.

Kosfeld, M., Heinrichis, M., and Zak, P.J. (2005). Oxytocin increases trust in humans. *Nature, 435*(7042), 673–676.

Kounios, J., and Beeman, M. (2009). The Aha! moment: The cognitive neuroscience of insight. *Current Directions in Psychological Science, 18*(4), 210–216.

Kounios, J., Fleck, J.I., and Green, D.L. et al. (2008). The origins of insight in resting-state brain activity. *Neuropsychologia, 46*(1), 281–291.

Kounios, J., Frymiare, J.L., and Bowden, F.M. et al. (2006). The prepared mind: Neural activity prior to problem presentation predicts subsequent solution by sudden insight. *Psychological Science, 17*(10), 882–890.

Kozlowski, S.W.J., and Klein, K.J. (2000). A multilevel approach to theory and research in organizations: Contextual, temporal, and emergent processes. In K.J. Klein and S.W.J. Kozlowski (eds), *Multilevel Theory, Research, and Methods in Organizations: Foundations, Extensions, and New Directions* (pp. 3–90). San Francisco, CA: Jossey-Bass.

Krain, A.L., Wilson, A.M., and Arbuckle, R. et al. (2006). Distinct neural mechanisms of risk and ambiguity: A meta-analysis. *NeuroImage, 32*(1), 477–484.

Kramer, R.M. (1999). Trust and distrust in organizations: Emerging perspectives, enduring questions. *Annual Review of Psychology, 50*, 569–598.

Kramer, R.M., and Tyler, T.R. (1996). *Trust in Organizations: Frontiers of Theory and Research*. Thousand Oaks, CA: Sage Publications.

Krawczyk, D.C. (2002). Contributions of the prefrontal cortex to the neural basis of human decision making. *Neuroscience and Behavioral Reviews*, 26(6), 631–664.

Krendl, A.C., Kensinger, E.A., and Ambady, N. (2012). How does the brain regulate negative bias to stigma? *Social Cognitive and Affective Neuroscience*, 7(6), 715–726.

Kreps, D.M. (1990). Corporate culture and economic behavior. In J. Alt and K. Shepsle (eds), *Perspective on Positive Political Economy* (pp. 90–143). Cambridge, UK: Cambridge University Press.

Kringelbach, M.L., and Berridge, K.C. (2016). Neuroscience of reward, motivation and drive. In S.I. Kim, J. Reeve, and M. Bong (eds), *Advances in Motivation and Achievement: Recent Developments in Neuroscience Research on Human Motivation* (pp. 23–35). Bingley, UK: Emerald Group Publishing Limited.

Krueger, F., Barbey, A.K., and McCabe, K. et al. (2009). The neural bases of key competencies of emotional intelligence. *Proceedings of the National Academy of Sciences*, 106(52), 22486–22491.

Krueger, N., and Welpe, I. (2014). Neuroentrepreneurship: What can entrepreneurship learn from neuroscience? In M. Morris (eds), *Annals of Entrepreneurship Education and Pedagogy* (pp. 60–90). Cheltenham, UK and Northampton, MA, USA: Edward Elgar Publishing.

Kühn, S., Ritter, S.M., and Müller, B.C.N. et al. (2013). The importance of the default mode network in creativity: A structural MRI study. *Journal of Creative Behavior*, 48(2), 152–163.

Lafferty, C.L., and Alford, K.L. (2010). Neuroleadership; Sustaining research relevance into the 21st century. *SAM Advanced Management Journal*, 75(3), 32–40.

Lakshminarayanan, V.R., and Santos, L.R. (2009). Cognitive preconditions for responses to fairness: An object retrieval test of inhibitory control in capuchin monkeys (*Cebus apella*). *Journal of Neuroscience, Psychology, and Economics*, 2(1), 12–20.

Lamichhane, B., Adhikari, B.M., Brosnan, S.F., and Dhamala, M. (2014). The neural basis of perceived unfairness in economic exchanges. *Brain Connectivity*, 4(8), 619–630.

Lammertsma, A.A. (1992). Position emission tomography. *Brain Topography*, 5(2), 113–117.

Lang, P.J., and Bradley, M.M. (2013). Appetitive and defensive motivation: Goal-directed or goal-determined? *Emotion Review*, 5(3), 230–234.

Lashley, K.S. (1930). Basic neural mechanisms in behavior. *Psychological Review*, 37(1), 1–24.

Latham, G.P., and Pinder, C.C. (2005). Work motivation theory and research at the dawn of the twenty-first century. *Annual Review of Psychology*, *56*, 485–516.

Lau, V., and Wong, Y. (2009). Direct and multiplicative effects of ethical dispositions and ethical climates on personal justice norms: A virtue ethics perspective. *Journal of Business Ethics*, *90*(2), 279–294.

Laureiro-Martinez, D., Brusoni, S., Canessa, N., and Zollo, M. (2015). Understanding the exploration–exploitation dilemma: An fMRI study of attention control and decision-making performance. *Strategic Management Journal*, *36*(3), 319–338.

Laureiro-Martinez, D., Brusoni, S., and Zollo, M. (2010). The neuroscientific foundations of the exploration–exploitation dilemma. *Journal of Neuroscience, Psychology, and Economics*, *3*(2), 95–115.

Laureiro-Martinez, D., Canessa, N., and Brusoni, S. et al. (2014). Frontopolar cortex and decision-making efficiency: Comparing brain activity of experts with different professional background during an exploration–exploitation task. *Frontiers in Human Neuroscience*, *7*(927), 1–10.

Lawler, III., E.E. (1973). Expectancy theory and job behavior. *Organizational Behavior and Human Performance*, *9*(3), 482–503.

Lawrence, P.R., and Nohria, N. (2002). *Driven: How Human Nature Shapes Our Choices.* Thousand Oaks, CA: Jossey-Bass.

Leavitt, K., Mitchell, T.R., and Peterson, J. (2010). Theory pruning: Strategies to reduce our dense theoretical landscape. *Organizational Research Methods*, *13*(4) 644–667.

LeBouc, R., and Pessiglione, M. (2013). Imaging social motivation: Distinct brain mechanisms drive effort production during collaboration versus competition. *Journal of Neuroscience*, *33*(40), 15894–15902.

LeDoux, J.E. (2000). Cognitive–emotional interactions: Listen to the brain. In R.D. Lane and L. Nadel (eds), *Cognitive Neuroscience of Emotion* (pp. 129–155). New York: Oxford University Press.

LeDoux, J.E. (2003). The emotional brain, fear, and the amygdala. *Cellular and Molecular Neurobiology*, *23*(4/5), 727–738.

Lee, D. (2008). Game theory and the neural basis of social decision making. *Neuroscience*, *11*(4), 404–409.

Lee, D., Seo, H., and Jung, M.W. (2012). Neural basis of reinforcement learning and decision making. *Annual Review of Neuroscience*, *35*, 287–308.

Lee, E.Y.J. (2015). The emotional link: Leadership and the role of implicit and explicit emotional contagion processes across multiple organizational levels. *Leadership Quarterly*, *26*(4), 654–670.

Lee, N., and Chamberlain, L. (2007). Neuroimaging and psychophysiological measurement in organizational research: An agenda for research in organizational cognitive neuroscience. *Annals of the New York Academy of Sciences*, *1118*, 18–42.

Lee, N., Senior, C., and Butler, M.J.R. (2012a). Leadership research and cognitive neuroscience: The state of this union. *The Leadership Quarterly*, *23*(2), 213–218.

Lee, N., Senior, C., and Butler, M.J.R. (2012b). The domain of organizational cognitive neuroscience: Theoretical and empirical challenges. *Journal of Management*, *38*(4), 921–934.

Lee, W. (2016). Insular cortex activity as the neural base of intrinsic motivation. In S.I. Kim, J. Reeve, and M. Bong (eds), *Advances in Motivation and Achievement: Recent Developments in Neuroscience Research on Human Motivation* (pp. 127–148). Bingley, UK: Emerald Group Publishing Limited.

Lee W., and Reeve J. (2012). Self-determined, but not non-self-determined, motivation predicts activations in the anterior insular cortex: An fMRI study of personal agency. *Social Cognitive and Affective Neuroscience*, *8*(5), 538–545.

Lee, W., Reeve J., Xue, Y., and Xiong J. (2012). Neural differences between intrinsic reasons for doing versus extrinsic reasons for doing: An fMRI study. *Neuroscience Research*, *73*(1), 68–72.

Legrenzi, P., and Umiltà, C. (2011). *Neuromania: On the Limits of Brain Science* [trans. Frances Anderson]. Oxford: Oxford University Press.

Lerner, J.S., Li, Y., Valdesolo, P., and Kassam, K.S. (2015). Emotion and decision making. *Annual Review of Psychology*, *66*, 799–823.

Leventhal G.S. (1980) What should be done with equity theory. In K.J. Gergen, M.S. Greenberg, and R.H. Willis (eds), *Social Exchange*. Boston, MA: Springer.

Lewicki, R., McAllister, D.J., and Bies, R.J. (1998). Trust and distrust: New relationships and realities. *Academy of Management Review*, *23*(3), 438–458.

Lewin, K. (1947). Frontiers in group dynamics: Concept, method and reality in social science: Social equilibria and social change. *Human Relations*, *1*(2), 143–153.

Li, W., Li, X., and Huang, L. (2015). Brain structure links trait creativity to openness to experience. *Social Cognitive and Affective Neuroscience*, *10*(2), 191–198.

Liberman, N., and Trope, Y. (1998). The role of feasibility and desirability considerations in near and distant future decisions: A test of temporal construal theory. *Journal of Personality and Social Psychology*, *75*(1), 5–18.

Lieberman, M.D. (2007a). Social cognitive neuroscience. In R.F. Baumeister and K.D. Vohs (eds), *Encyclopedia of Social Psychology*. Thousand Oaks, CA: Sage.

Lieberman, M.D. (2007b). Social cognitive neuroscience: A review of core processes. *Annual Review of Psychology*, *58*, 259–289.

Lieberman, M.D., Eisenberger, N.I., and Crockett, M.J. et al. (2007). Putting feelings into words: Affect labeling disrupts amygdala activity in response to affective stimuli. *Psychological Science*, *18*(5), 421–428.

Lieberman, M.D., Gaunt, R., Gilbert, D.T., and Trope, Y. (2002). Reflexion and reflection: A social cognitive neuroscience approach to attributional influence. In M.P. Zanna (ed.), *Advances in Experimental Social Psychology* (Vol. 34, pp. 199–249). New York: Academic Press.

Lind, E.A. (2001). Fairness heuristics theory: Justice judgments as pivotal cognitions in organizational relations. In J. Greenberg and R. Cropanzano (eds), *Advances in Organizational Justice* (pp. 56–88). Stanford, CA: Stanford University Press.

Lind, E.A., and Van den Bos, K. (2002). When fairness works: Toward a general theory of uncertainty management. In B.M. Staw and R.M. Kramer (eds), *Research in Organizational Behavior* (Vol. 24, pp. 181–223). Greenwich, CT: JAI Press.

Lindebaum, D. (2013a). Ethics and the neuroscientific study of leadership: A synthesis and rejoinder to Ashkanasy, Cropanzano and Becker, and McLagan. *Journal of Management Inquiry*, *22*(3), 317–323.

Lindebaum, D. (2013b). Pathologizing the healthy but ineffective: Some ethical reflections on using neuroscience in leadership research. *Journal of Management Inquiry, 22(3)*, 295–305.

Lindebaum, D. (2016). Critical essay: Building new management theories on sound data? The case of neuroscience. *Human Relations*, *69*(3), 537–550.

Lindebaum, D., Al-Ahmoudi, I., and Brown, V.L. (2017). Does leadership development need neuroethics? *Academy of Management Learning and Education*, in press.

Lindebaum, D., and Jordan, P.J. (2014). A critique on neuroscientific methodologies in organizational behavior and management studies. *Journal of Organizational Behavior*, *35*(7), 898–908.

Lindebaum, D., Geddes, D., and Gabriel, Y. (2016). Moral emotions and ethics in organizations: Introduction to the special issue. *Journal of Business Ethics*, *141*(4), 645–656.

Lindebaum, D., and Zundel, M. (2013). Not quite a revolution: Scrutinizing organizational neuroscience in leadership studies. *Human Relations*, *66*(6), 857–877.

Locke, E.A. (1996). Motivation through conscious goal setting. *Applied and Preventive Psychology*, *5*(2), 117–124.

Locke, E.A., and Latham, G.P. (1990). *A Theory of Goal Setting and Task Performance*. Englewood Cliffs, NJ: Prentice-Hall.

Locke, E.A., and Latham, G.P. (2002). Building a practically useful theory of goal setting and task motivation: A 35-year odyssey. *American Psychologist*, *57*(9), 705–717.

Locke, E.A., and Latham, G.P. (2006). New directions in goal setting theory. *Current Directions in Psychological Science*, *15*(5), 265–268.

Loewenstein, G., Rick, S., and Cohen, J.D. (2008). Neuroeconomics. *Annual Review of Psychology*, *59*, 647–672.

Lord, R.G., Diefendorff, J.M., Schmidt, A.M., and Hall, R.J. (2010). Self-regulation at work. *Annual Review of Psychology*, *61*, 543–568.

Lorensen, T.D., and Dickson, P. (2003). Quantitative EEG normative databases: A comparative investigation. *Journal of Neurotherapy*, *7*(3/4), 53–68.

Luck, S.J. (2005). *An Introduction to the Event-related Potential Technique*, Cambridge, MA: MIT Press.

Luo, J., Li, W., and Qiu, J. et al. (2013). Neural basis of scientific innovation induced by heuristic prototype. *PLOS One*, *8*(1), 1–7.

Luo, Q., Nakic, M., and Wheatley, T. et al. (2006). The neural basis of implicit moral attitude: An IAT study using event-related fMRI. *NeuroImage*, *30*(4), 1449–1457.

Ma, Q., Jin, J., Meng, L., and Shen, Q. (2014). The dark side of monetary incentive: How does extrinsic reward crowd out intrinsic motivation. *NeuroReport*, *25*(3), 194–198.

Ma, W.J., and Jazayeri, M. (2014). Neural coding of uncertainty and probability. *Annual Review of Neuroscience*, *37*, 205–220.

Malpass, R.S., and Kravitz, J. (1969). Recognition for faces of own and other race. *Journal of Personality and Social Psychology*, *13*(4), 330–334.

Mandler, G. (2002). Origins of the cognitive revolution. *Journal of the History of the Behavioral Sciences*, *38*(4), 339–353.

March, J.G. (1991). Exploration and exploitation in organizational learning. *Organization Science*, *2*(1), 71–87.

Marcus, S.J. (2002). *Neuroethics: Mapping the Field*. New York: The Dana Foundation.

Marsden, K.E., Ma, W.J., and Deci, E.L. et al. (2015). Diminished neural responses predict enhanced intrinsic motivation and sensitivity to external incentive. *Cognitive and Affective Behavioral Neuroscience*, *15*(2), 276–286.

Marshall, P.J. (2009). Relating psychology and neuroscience. *Perspectives on Psychological Science*, *4*(2), 113–125.

Martindale, C. (1999). Biological basis of creativity. In R.J. Sternberg (ed.), *Handbook of Creativity* (pp. 137–152). Cambridge, UK: Cambridge University Press.

Martins, N. (2010). Can neuroscience inform neuroeconomics? Rationality, emotions and preference formation. *Cambridge Journal of Economics, 35*(2), 251–267.

Maslow, A.H. (1954). *Motivation and Personality*. New York: Harper & Row.

Mathieu, J.E., Heffner, T.S., and Goodwin, G.F. et al. (2000). The influence of shared mental models on team process and performance. *Journal of Applied Psychology, 85*(2), 273–283.

Mathur, V.A., Harada, T., Lipke, T., and Chiao, J.Y. (2010). Neural basis of extraordinary empathy and altruistic motivation. *NeuroImage, 51*(4), 1468–1475.

Mayhew, S.D., Ostwald, D., Porcaro, C., and Bagshaw, A.P. (2013). Spontaneous EEG alpha oscillation interacts with positive and negative BOLD responses in the visual-auditory cortices and default mode network. *NeuroImage, 76*(August), 362–372.

McAllister, D.J. (1995). Affect- and cognition-based trust as foundations for interpersonal cooperation in organizations. *Academy of Management Journal, 38*(1), 24–59.

McCabe, D., and Castel, A. (2008). Seeing is believing: The effect of brain images on judgments of scientific reasoning. *Cognition, 107*(1), 343–352.

McCabe, K., Houser, D., and Ryan, L. et al. (2001). A functional imaging study of cooperation in two-person reciprocal exchange. *Proceedings of the National Academy of Sciences, 98*(20), 11832–11835.

McClelland, D.C. (1961). *The Achieving Society*. Princeton, NJ: Van Nostrand.

McClelland, D.C. (1965). Need for achievement and entrepreneurship: A longitudinal study. *Journal of Personality and Social Psychology, 1*(4), 389–392.

McClure, S.M., York, M.K., and Montague, P.R. (2004). The neural substrates of reward processing in humans. The modern role of fMRI. *The Neuroscientist, 10*(3), 260–268.

Mednick, S.A. (1962). The associative basis of the creative process. *Psychological Review, 69*(3), 220–232.

Meng, L., and Ma, Q. (2015). Live as we choose: The role of autonomy in facilitating intrinsic motivation. *International Journal of Psychophysiology, 98*(3), 441–447.

Metuki, N., Sela, T., and Levador, M. (2012). Enhancing cognitive control components of insight problem solving by anodal tDCS of the left dorsolateral prefrontal cortex. *Brain Stimulation, 5*(2), 110–115.

Michel, C.M., and Murray, M.M. (2012). Towards the utilization of EEG as a brain imaging tool. *NeuroImage, 61*(2), 371–385.

Miller, E.K., and Cohen, J.D. (2001). An integrative theory of prefrontal cortex function. *Annual Review of Neuroscience, 24,* 167–202.

Miller, E.M., Shankar, M.U., Knutson, B., and McClure, S.M. (2014). Dissociating motivation from reward in human striatal activity. *Journal of Cognitive Neuroscience, 26*(5), 1075–1084.

Miller, G.A. (1994). The magical number seven, plus or minus two: Some limits on our capacity for processing information. *Psychological Bulletin, 101*(2), 343–352.

Miller, G.A. (2003). The cognitive revolution: A cognitive perspective. *Trends in Cognitive Sciences, 7*(3), 141–144.

Mintzberg, H. (1976). Planning on the left and managing on the right side. *Harvard Business Review, 54,* 49–58.

Mintzberg, H. (1979). *The Structuration of Organizations.* Upper Saddle River, NJ: Pearson.

Mitchell, J.P., Macrae, C.N., and Banaji, M.R. (2006). Dissociable medial prefrontal contributions to judgments of similar and dissimilar others. *Neuron, 50*(4), 655–663.

Mizuno, K., Tanaka, M., and Ishii, A. (2008). The neural basis of academic achievement motivation. *NeuroImage, 42*(1), 369–378.

Mohammed, S., and Dumville, B. (2001). Team mental models in a team knowledge framework: Expanding theory and measurement across disciplinary boundaries. *Journal of Organizational Behavior, 22*(2), 89–106.

Molenberghs, P., Prochilo, G., and Steffens, N.K. et al. (2015). The neuroscience of inspirational leadership: The importance of collective-oriented language and shared group membership. *Journal of Management.* Doi.org/10.1177/0149206314565242.

Moll, J., de Oliveira-Souza, R.A. and Eslinger, P.J. et al. (2002). The neural correlates of moral sensitivity: A functional magnetic resonance imaging investigation of basic and moral emotions. *Journal of Neuroscience, 22*(7), 2730–2736.

Moll, J., de Oliveira-Souza, R.A., and Eslinger, P.J. (2003). Morals and the human brain: A working model. *NeuroReport, 14*(3), 299–305.

Moll, J., de Oliveira-Souza, R.A., and Zahn, R. (2008). The neural basis of moral cognition: Sentiments, concepts, and values. *Annals of the New York Academy of Sciences, 1124,* 161–180.

Moll, J., Zahn, R., and De Oliveira-Souza, R. et al. (2005). Opinion: The neural basis of human moral cognition. *Nature Reviews: Neuroscience, 6*(10), 799–809.

Montada, L., and Schneider, A. (1989). Justice and emotional reactions to the disadvantaged. *Social Justice Research, 3*(4), 313–344.

Montague, P.R., Lohrenz, T., and Dayan, P. (2015). The three R's of trust. *Current Opinion in Behavioral Sciences, 3,* 102–106.

Moors, A., and De Houwer, J. (2006). Automaticity: A conceptual and theoretical analysis. *Psychological Bulletin*, *132*(2), 297–326.

Moretto, G., Ladavas, E., Mattioli, F., and Di Pellegrino, G. (2010). A psychophysiological investigation of moral judgment after ventromedial prefrontal damage. *Journal of Cognitive Neuroscience*, *22*(8), 1888–1899.

Morse, G. (2006). Decisions and desire. *Harvard Business Review*, January, 42–51.

Mukamel, R., and Fried, I. (2012). Human intracranial recording and cognitive neuroscience. *Annual Review of Psychology*, *63*, 511–537.

Murayama, K., Matsumoto, M., Izuma, K., and Matzumoto, K. (2010). Neural basis of the undermining effect of monetary reward on intrinsic motivation. *Proceedings of the National Academy of Sciences*, *107*(49), 20911–20916.

Murayama, K., Matsumoto, M., and Izuma, K. et al. (2015). How self-determined choice facilitates performance: A key role of the ventromedial prefrontal cortex. *Cerebral Cortex*, *25*(5), 1241–1251.

Naqvi, S., Shiv, B., and Bechara, A. (2006). The role of emotion in decision making: A cognitive neuroscience perspective. *Current Directions in Psychological Science*, *15*(5), 260–264.

Nash, K., Baumgartner, T., and Knoch, D. (2017). Group-focused morality is associated with limited conflict detection and resolution capacity: Neuroanatomical evidence. *Biological Psychology*, *123*(February), 235–240.

Nave, G., Camerer, C., and McCullough, M. (2015). Does oxytocin increase trust in humans? A critical review of research. *Perspectives on Psychological Science*, *10*(6), 772–789.

Neisser, U. (1967). *Cognitive Psychology*. New York: Psychology Press.

Newell, A., and Simon, H.A. (1976). Computer science as empirical enquiry. *Communications of the ACM*, *19*(3), 113–126.

Nicholson, N. (1998). How hardwired is human behavior? *Harvard Business Review*, July–August, 136–147.

Nicolaou, N., and Shane, S. (2014). Biology, neuroscience, and entrepreneurship. *Journal of Management Inquiry*, *23*(1), 98–100.

Niv, Y., Joel, D., and Dayan, P. (2006). A normative perspective on motivation. *Trends in Cognitive Sciences*, *10*(8), 375–381.

Nohria, N., Groysberg, B., and Lee, L.E. (2008). Employee motivation. *Harvard Business Review* (July–August), 1–8.

Nowak, M.A. (2006). Five rules for the evolution of cooperation. *Science*, *314*(5805), 1560–1563.

Nowak, M.A., Page, K.M., and Sigmund, K. (2000). Fairness versus reason in the ultimatum game. *Science*, *289*(5485), 1773–1775.

Oatley, K., and Johnson-Laird, P. (1987). Toward a cognitive theory of emotions. *Cognition and Emotion, 1*(1), 29–50.

Ochsner, K.N., and Gross, J.J. (2005). The cognitive control of emotion. *Trends in Cognitive Sciences, 9*(5), 242–249.

Ochsner, K.N., and Lieberman, M.D. (2001). The emergence of social cognitive neuroscience. *American Psychologist, 56*(9), 717–734.

O'Connor, C., Rees, G., and Joffe, H. (2012). Neuroscience in the public sphere. *Neuron, 74*(2), 220–226.

O'Doherty, J.P., Cockburn, J., and Pauli, W.M. (2016). Learning, reward, and decision making. *Annual Review of Psychology, 68*, 73–100.

Öhman, A. (2005). The role of the amygdala in human fear: Automatic detection of threat. *Psychoneuroendocrinology, 30*(10), 953–958.

Ollinger, J.M., and Fessler, J.A. (1997). Position emission tomography. *IEEE Signal Processing Magazine*, August, 43–55.

Olteanu, M.D.B. (2015). Neuroethics and responsibility in conducting neuromarketing research. *Neuroethics, 8*(2), 191–202.

Onarheim, B., and Friis-Olivarius, M. (2013). Applying the neuroscience of creativity to creativity training. *Frontiers in Human Neuroscience, 7*(October), 1–10.

Orlitzky, M. (2017). How cognitive neuroscience informs a subjectivist-evolutionary explanation of business ethics. *Journal of Business Ethics, 144*(4), 717–732.

Partridge, B.J., Bell, S.K., and Lucke, J.C. et al. (2011). Smart drugs "as common as coffee": Media hype about neuroenhancement. *PLOS One, 6*(11), 1–8.

Pascual-Leone, A., Walsh, V., and Rothwell, J. (2000). Transcranial magnetic stimulation in cognitive neuroscience-virtual lesion, chronometry, and functional connectivity. *Current Opinion in Neurobiology, 10*(2), 232–237.

Paulus, M.P., Potterat, E.G., and Taylor, M.K. et al. (2009). A neuroscience approach to optimizing brain resources for human performance in extreme environments. *Neuroscience and Bio-Behavioral Reviews, 33*(7), 1080–1088.

Peterson, S.J., Balthazard, P.A., Waldman, D.A., and Thatcher, R.W. (2008). Neuroscientific implications of psychological contract: Are the brains of optimistic, hopeful, confident, and resilient leaders different? *Organizational Dynamics, 37*(4), 342–353.

Pfenninger, K.H., and Shubik, V.R. (2001). Insights into the foundations of creativity: A synthesis. In K.H. Pfenninger and V.R. Shubik (eds), *The Origins of Creativity* (pp. 213–236). Oxford: Oxford University Press.

Phan, K.L., Taylor, S.F., and Welsh, R.C. (2004). Neural correlates of individual ratings of emotional salience: A trial-related fMRI study. *NeuroImage, 21*(2), 768–780.

Phelps, E.A. (2001). Faces and races in the brain. *Nature Neuroscience, 4*(8), 775–776.

Phelps, E.A. (2006). Emotion and cognition: Insights from studies of the human amygdala. *Annual Review of Psychology, 57*, 27–53.

Phelps, E.A., Cannistraci, C.J., and Cunningham, W.A. (2003). Intact performance on an indirect measure of race bias following amygdala damage. *Neuropsychologia, 41*(2), 203–208.

Phelps, E.A., Lempert, K.M., and Sokol-Hessner, P. (2014). Emotion and decision-making: Multiple modulatory circuits. *Annual Review of Neuroscience, 37*, 263–287.

Phelps, E.A., O'Connor, K.J., and Cunningham, W.A. et al. (2000). Performance on indirect measures of race evaluation predicts amygdala activation. *Journal of Cognitive Neuroscience, 12*(5), 729–738.

Phillip, V.L. (1985). Defining business ethics: Like nailing jello to a wall. *Journal of Business Ethics, 4*(5), 377–383.

Pieritz, A.K., Thybusch, K., and Rutter, B. et al. (2012). Creativity and the brain: Uncovering the neural signature of conceptual expansion. *Neuropsychologia, 50*(8), 1906–1917.

Pillutla, M.M., and Murnighan, J.K. (1996). Unfairness, anger, and spite: Emotional rejections of ultimatum offers. *Organizational Behavior and Human Decision Processes, 68*(3), 208–224.

Pinder, C.C. (1998). *Work Motivation in Organizational Behavior.* Upper Saddle River, NJ: Prentice Hall.

Platt, M.L., and Huettel, S.A. (2008). Risky business: The neuroeconomics of decision making under uncertainty. *Nature Neuroscience, 11*(4), 398–403.

Poldrack, R.A. (2006). Can cognitive processes be inferred from neuroimaging data? *Trends in Cognitive Sciences, 10*(2), 59–63.

Poldrack, R.A. (2008). The role of fMRI in cognitive neuroscience: Where do we stand? *Current Opinion in Neurobiology, 18*(2), 223–227.

Poreisz, C., Boros, K., Antal, A., and Paulus, W. (2007). Safety aspects of transcranial direct current stimulation concerning healthy subjects and patients. *Brain Research Bulletin, 72*(4), 208–214.

Powell, T.C. (2011). Neurostrategy. *Strategic Management Journal, 32*(3), 1484–1499.

Powell, T.C., and Puccinelli, N.M. (2012). The brain as substitute for strategic organization. *Strategic Organization, 10*(3), 207–214.

Prehn K., Wartenburger, I., and Meriau, K. et al. (2008). Individual differences in moral judgment competence influence neural correlates of socio-normative judgments. *Social Cognitive and Affective Neuroscience*, *3*(1), 33–46.

Premack, D., and Woodruff, G. (1978). Does the chimpanzee have a theory of mind? *Behavioral and Brain Sciences*, *1*(4), 515–526.

Pressman, S. (2006). Kahneman, Tversky, and institutional economics. *Journal of Economic Issues*, *XL*(2), 501–506.

Preston, S.D., and De Waal, F.B.M. (2002). Empathy: Its ultimate and proximal bases. *The Behavioral and Brain Sciences*, *25*(1), 1–72.

Pugh, S.D., Groth, M., and Hennig-Thurau, T. (2011). Willing and able to fake emotions: A closer examination of the link between emotional dissonance and employee well-being. *Journal of Applied Psychology*, *96*(2), 377–390.

Quirin, M., Meyer, F., and Heise, N. et al. (2013). Neural correlates of social motivation: An fMRI study on power versus affiliation. *International Journal of Psychophysiology*, *88*(3), 289–295.

Rabin, M. (1993). Incorporating fairness into game theory and economics. *American Economic Review*, *83*(5), 1281–1302.

Raichle, M.E., MacLeod, A.M., and Snyder, A.Z. (2001). A default mode of brain function. *Proceedings of the National Academy of Sciences*, *98*, 676–682.

Raihani, N.J., and McAuliffe, K. (2012). Human punishment is motivated by inequity aversion, not a desire for reciprocity. *Biology Letters*, *8*(5), 802–804.

Ramchandran, K., Colbert, A.E., and Brown, K.G. (2016). Exploring the neuropsychological antecedents of transformational leadership: The role of executive function. *Adaptive Human Behavior and Physiology*, *2*(4), 325–343.

Rameson, L.T., Morelli, S.A., and Lieberman, M.D. (2011). The neural correlates of empathy: Experience, automaticity, and prosocial behavior. *Journal of Cognitive Neuroscience*, *24*(1), 235–245.

Rangel, A., and Hare, T. (2010). Neural computations associated with goal-directed choice. *Current Opinion in Neurobiology*, *20*(2), 262–270.

Rangel, A., Camerer, C., and Montague, P.R. (2008). A framework for studying the neurobiology of value-based decision-making. *Nature Reviews Neuroscience*, *9*, 545–556.

Reina, C.S., Peterson, S.J., and Waldman, D.A. (2015). Neuroscience as a basis for understanding emotions and affect in organizations. In D.A. Waldman and P.A. Balthazard (eds), *Organizational Neuroscience (Monographs in Leadership and Management* (Vol. 7, pp. 213–232). Bingley, UK: Emerald Group Publishing Limited.

Reinharth, L., and Wahba, M.A. (1975). Expectancy theory as a predictor of work motivation, effort expenditure, and job performance. *Academy of Management Journal*, *18*(3), 520–537.

Rest, J.R. (1984). The major components of morality. In W. Kurtines and J. Gewirtz (eds), *Morality, Moral Behavior, and Moral Development* (pp. 24–38). New York: John Wiley and Sons.

Rest, J.R., Bebeau, M.J., and Volker, J. (1986). An overview of the psychology of morality. In J.R. Rest (ed.), *Moral Development: Advances in Research and Theory* (pp. 1–39). Boston, MA: Praeger.

Rest, J.R., Narvaez, D., Bebeau, M.J., and Thomas, S.J. (1999). *Postconventional Moral Thinking: A Neo-Kohlbergian Approach*. Mahwah, NJ: Erlbaum.

Rettinger, D.A., and Hastie, R. (2003). Comprehension and decision making. In S.L. Schneider, and J. Shanteau (eds), *Emerging Perspectives on Judgment and Decision Research: Cambridge Series on Judgment and Decision Making* (pp. 165–200). New York: Cambridge University Press.

Reynolds, S.J. (2006). A neurocognitive model of the ethical decision-making process: Implications for study and practice. *Journal of Applied Psychology*, *91*(4), 737–748.

Reynolds, S.J. (2008). Moral attentiveness: Who pays attention to the moral aspects of life? *Journal of Applied Psychology*, *93*(5), 1027–1041.

Richeson, J.A., Baird, A.A., and Gordon, H.L. et al. (2003). An fMRI investigation of the impact of interracial contact on executive function. *Nature Neuroscience*, *6*(12), 1323–1328.

Rick, S. (2011). Losses, gains, and brains: Neuroeconomics can help to answer open questions about loss aversion. *Journal of Consumer Psychology*, *21*(4), 453–463.

Riedl, R., and Javor, A. (2012). The biology of trust: Integrating evidence from genetics, endocrinology, and functional brain imaging. *Journal of Neuroscience, Psychology, and Economics*, *5*(2), 63–91.

Rilling, J.K., and Sanfey, A.G. (2011). The neuroscience of social decision making. *Annual Review of Psychology*, *62*, 23–48.

Rilling, J.K., Goldsmith, D.R., and Glenn, A.L. (2008). The neural correlates of the affective response to unreciprocated cooperation. *Neuropsychologia*, *46*(5), 1256–1266.

Rilling, J.K., Gutman, D.A., and Zeh, T.R. et al. (2002). A neural basis for social cooperation. *Neuron*, *35*(2), 395–405.

Rilling, J.K., King-Casas, B., and Sanfey, A.G. (2008). The neurobiology of social decision-making. *Current Opinion in Neurobiology*, *18*(2), 159–165.

Rilling, J.K., Sanfey, A.G., and Aronson, J.A. (2004a). Opposing BOLD responses to reciprocated and unreciprocated altruism in putative reward pathways. *NeuroReport, 15*(16), 1–5.

Rilling, J.K., Sanfey, A.G., and Aronson, J.A. et al. (2004b). The neural correlates of theory of mind within interpersonal interactions. *Neuro-Image, 22*(4), 1694–1703.

Riolo, R.L., Cohen, M.D., and Axerold, R. (2001). Evolution of cooperation without reciprocity. *Nature, 414*, 441–443.

Rizzolatti, G., and Craighero, L. (2004). The mirror-neuron system. *Annual Review of Neuroscience, 27*, 169–192.

Rizzolatti, G., and Fabbri-Destro, M. (2008). The mirror system and its role in social cognition. *Current Opinion in Neurobiology, 18*(2), 179–184.

Robbins, T.W., and Everitt, B.J. (1996). Neurobiological mechanisms of reward and motivation. *Current Opinion in Neurobiology, 6*(2), 228–236.

Roberson, Q., Holmes IV, O., and Perry, J.L. (2017). Transforming research on diversity and firm performance: A dynamic capabilities perspective. *Academy of Management Annals, 11*(1), 189–216.

Robertson, D., Snarey, J., and Ousley, A. et al. (2007). The neural processing of moral sensitivity to issues of justice and care. *Neuropsychologia, 45*(8), 755–766.

Robertson, D.C., Voegtlin, C., and Maak, T. (2017). Business ethics: The promise of neuroscience. *Journal of Business Ethics, 144*(4), 679–697.

Robinson, S.L. (1996). Trust and the breach of the psychological contract. *Administrative Science Quarterly, 41*(4), 574–599.

Rochford, K.C., Jack, A.I., Boyatzis, R.E., and French, S.E. (2017). Ethical leadership as a balance between opposing neural networks. *Journal of Business Ethics, 144*(4), 755–770.

Roskies, A. (2002). Neuroethics for the new millennium. *Neuron, 35*(1), 21–23.

Runco, M.A. (2004). Creativity. *Annual Review of Psychology, 55*, 657–687.

Runco, M.A., and Jaeger, G.J. (2012). The standard definition of creativity. *Creativity Research Journal, 24*(1), 92–96.

Ryan, L.V. (2017). Sex differences through a neuroscience lens: Implications for business ethics. *Journal of Business Ethics, 144*(4), 771–782.

Ryan R.M., and Deci, E.L. (2000). Self-determination theory and the facilitation of intrinsic motivation, social development, and well–being. *American Psychologist, 55*(1), 68–78.

Rypma, B., Berger, V., and Prabhakaran, V. et al. (2006). Neural correlates of cognitive efficiency. *NeuroImage, 33*(3), 969–979.

Sacco, D.F., Brown, M., Lustgraaf, C.J.N., and Hugenberg, K. (2017). The adaptive utility of deontology: Deontological moral decision-making fosters perceptions of trust and likeability. *Evolutionary Psychological Science*, *3*(2), 125–132.

Salomone, J.D. (1994). The involvement of nucleus accumbens dopamine and aversive motivation. *Behavioural Brain Research*, *61*(2), 117–133.

Salvador, R., and Folger, R.G. (2009). Business ethics and the brain. *Business Ethics Quarterly*, *19*(1), 1–30.

Sandkühler, S., and Bhattacharya, J. (2008). Deconstructing insight: EEG correlates of insightful problem solving. *PLOS One*, *3*(1), 1–12.

Sanfey, A.G. (2007). Social decision-making: Insights from game theory and neuroscience. *Science*, *318*(5850), 598–602.

Sanfey, A.G., Loewenstein, G., McClure, S.M., and Cohen, J.D. (2006). Neuroeconomics: Cross-currents in research on decision-making. *Trends in Cognitive Sciences*, *10*(3), 108–116.

Sanfey, A.G., Rilling, J.K., Aronson, and L.E., Nystrom (2003). The neural basis of economic decision-making in the ultimate game. *Science*, *300*(5626), 1755–1758.

Sanfey, A.G., Stallen, M., and Chang, L.J. (2014). Norms and expectations in social decision making. *Trends in Cognitive Sciences*, *18*(4), 172–174.

Santos, L.R., and Rosati, A.G. (2015). The evolutionary roots of human decision making. *Annual Review of Psychology*, *66*, 321–347.

Sarasvathy, S.D. (2001). Causation and effectuation: Toward a theoretical shift from economic inevitability to entrepreneurial contingency. *Academy of Management Review*, *26*(2), 243–288.

Satpute, A.B., and Lieberman, M.D. (2006). Integrating automatic and controlled processes into neurocognitive models of social cognition. *Brain Research*, *1079*(1), 86–97.

Sawyer, R. (2011). The cognitive neuroscience of creativity: A critical review. *Creativity Research Journal*, *23*(2), 137–154.

Schjoedt, U., Stødkilde–Jørgensen, H., and Geertz, A.W. et al. (2010). The power of charisma – perceived charisma inhibits the frontal executive network of believers in intercessory prayer. *Social Cognitive and Affective Neuroscience*, *6*(1), 119–127.

Schmitt, M.J., Neumann, R., and Montada, L. (1995). Dispositional sensitivity to befallen injustice. *Social Justice Research*, *8*(4), 385–407.

Schultz, W. (2002). Getting formal with dopamine and reward. *Neuron*, *36*(2), 241–263.

Scott, G., Leritz, L.E., and Mumford, M.D. (2004). The effectiveness of creativity training: A quantitative review. *Creative Research Journal*, *16*(4), 361–388.

Seeley, W.W., Menon, V., and Schatzberg, A.F. et al. (2007). Dissociable intrinsic connectivity networks for salience processing and executive control. *Journal of Neuroscience*, *27*(9), 2349–2356.

Selye, H. (1956). *The Stress of Life*. New York: McGraw-Hill.

Senior, C., Lee, N., and Butler, M.J.R. (2011). Organizational cognitive neuroscience. *Organization Science*, *22*(3), 804–815.

Sent, E.M. (2004). Behavioral economics: How psychology made its (limited) way back into economics. *History of Political Economy*, *36*(4), 735–760.

Sessa, P., Tomelleri, S., and Luria, R. et al. (2012). Look out for strangers: Sustained neural activity during visual working memory maintenance of other-race faces is modulated by implicit racial prejudice. *Social Cognitive and Affective Neuroscience*, *7*(3), 314–321.

Seymour, B., Singer, T., and Dolan, R. (2007). The neurobiology of punishment. *Nature Reviews Neuroscience*, *8*(4), 300–311.

Shamay-Tsoory, S.G., Adler, N., and Aharon-Peretz, J. et al. (2011). The origins of originality: The neural bases of creative thinking and originality. *Neuropsychologia*, *49*(2), 178–185.

Shamir, B., House, R.J., and Arthur, M.B. (1993). The motivational effects of charismatic leadership: A self-concept based theory. *Organization Science*, *4*(4), 577–594.

Simon, H.A. (1947). *Administrative Science*. New York: Macmillan.

Simon, H.A. (1955). A behavioral theory of rational choice. *Quarterly Journal of Economics*, *69*(1), 99–118.

Simon, H.A. (1956). Rational choice and the structure of the environment. *Psychological Review*, *63*(2), 129–138.

Simon, H.A. (1965). The new science of management decision. In H.A. Simon (ed.), *The Shape of Automation for Men and Management* (pp. 51–79). New York: Harper & Row.

Simon, H.A. (1977). *Models of Discovery*. Dordrecht: Reidel Publishing Company.

Simon, H.A. (1980). *The Sciences of the Artificial*. Cambridge, MA: MIT Press.

Singer, T., Seymour B., and O'Doherty et al. (2006). Empathic neural responses are modulated by the perceived fairness of others. *Nature*, *439*(7075), 466–469.

Slingerland, E. (2011). Of what use are the odes? Cognitive science, virtue ethics, and early Confucian ethics. *Philosophy East and West*, *61*(1), 80–109.

Smallwood, J., Brown, K., Baird, B., and Schooler, J.W. (2012). Cooperation between the default mode network and the fronto-parietal network in the production of an internal train of thought. *Brain Research*, *1428*(January), 60–70.

Smetana, J.G., and Killen, M. (2008). Moral cognition, emotions, and neuroscience: An integrative developmental view. *European Journal of Developmental Science*, *2*(3), 324–339.

Smith, E.R., and Semin, G.R. (2004). Socially situated cognition: Cognition in its social context. In M.P. Zanna (ed.), *Advances in Experimental Social Psychology* (Vol. 36, pp. 53–117). San Diego, CA: Academic Press.

Smith, R., and Lane, R.D. (2015). The neural basis of one's own conscious and unconscious emotional states. *Neuroscience & Biobehavioral Reviews*, *57*, 1–29.

Sowden, P.T., Pringle, A., and Gabora, L. (2015). The shifting sands of creative thinking: Connections to dual-process theory. *Thinking and Reasoning*, *21*(1), 40–60.

Spence, C. (2016). Neuroscience-inspired design: From academic neuromarketing to commercially relevant research. *Organizational Research Methods*. Doi.org/10.1177/1094428116672003.

Spence, M. (1973). Job market signaling. *Quarterly Journal of Economics*, *87*(3), 355–374.

Sperry, R.W. (1993). The impact and promise of the cognitive revolution. *American Psychologist*, *48*(8), 878–885.

Sporer, S.L. (2001). Recognizing faces of other ethnic groups: An integration of theories. *Psychology, Public Policy, and Law*, *7*(1), 36–97.

Stallen, M., and Sanfey, A.G. (2013). The cooperative brain. *The Neuroscientist*, *19*(3), 292–303.

Stam, C.J. (2010). Use of magnetoencephalography (MEG) to study functional brain networks in neurodegenerative disorders. *Journal of Neurological Sciences*, *289*(1/2), 128–134.

Stein, M.I. (1953). Creativity and culture. *Journal of Psychology*, *36*(2), 311–322.

Sternberg, R.J. (1981). Testing and cognitive psychology. *American Psychologist*, *36*(10), 1181–1189.

Sternberg, R.J., and Davidson, J.E. (eds) (1995). *The Nature of Insight*. Cambridge, MA: MIT Press.

Sternberg, R.J., and Lubart, T.I. (1996). Investing in creativity. *American Psychologist*, *51*(7), 677–688.

Stevens, J.R., and Stephens, D.W. (2004). The economic basis of cooperation: Tradeoffs between selfishness and generosity. *Behavioral Ecology*, *15*(2), 255–261.

Stogdill, R.M. (1948). Personal factors associated with leadership: A survey of the literature. *Journal of Psychology*, *25*(1), 35–71.

Strobel, A., Zimmermann, J., and Schmitz, A. et al. (2011). Beyond revenge: Neural and genetic bases of altruistic punishment. *Neuro-Image, 54*(1), 671–680.

Suhler, C.L., and Churchland, P. (2011). Can innate, modular foundations explain morality? Challenges for Haidt's moral foundations theory. *Journal of Cognitive Neuroscience, 23*(9), 2103–2116.

Sun, J., Chen, Q., and Zhang, Q. et al. (2016). Training your brain to be more creative: Brain functional and structural changes induced by divergent thinking training. *Human Brain Mapping, 37*(10), 3375–3387.

Suzuki, S., Niki, K., Fujisaki, S., and Akiyama, E. (2011). Neural basis of conditional cooperation. *Social Cognitive and Affective Neuroscience, 6*(3), 338–347.

Tabibnia, G., and Lieberman, M.D. (2007). Fairness and cooperation are rewarding: Evidence from social cognitive neuroscience. In C. Senior and M.J.R. Butler (eds), *The Social Cognitive Neuroscience of Organizations. Annals of the New York Academy of Sciences* (Vol. 118, pp. 90–101). Boston, MA: Blackwell Publishing.

Tabibnia, G., Satpute, A.B., and Lieberman, M.D. (2008). The sunny side of fairness: Preference for fairness activates reward circuitry (and disregarding unfairness activates self-control circuitry). *Psychological Science, 19*(4), 339–347.

Tajfel, H. (1970). Experiments in intergroup discrimination. *Scientific American, 223*(5), 96–102.

Tajfel, H., and Turner, J.C. (1979). An integrative theory of intergroup conflict. In W.G. Austin and S. Worchel (eds), *The Social Psychology of Intergroup Relations* (pp. 33–47). Monterey, CA: Brooks/Cole.

Takeuchi, H., Taki, Y., and Nouchi, R. et al. (2014). Regional gray matter density is associated with achievement motivation: Evidence from voxel-based morphometry. *Brain Structure and Function, 219*(1), 71–83.

Takeuchi, H., Taki, Y., and Sassa, Y. et al. (2010). White matter structures associated with creativity: Evidence from diffusion tensor imaging. *NeuroImage, 51*(1), 11–18.

Thaler, R.H. (1980). Toward a theory of consumer choice. *Journal of Economic Behavior and Organization, 1*(1), 39–60.

Thaler, R.H. (1988). Anomalies: The ultimatum game. *Journal of Economic Perspectives, 2*(4), 195–206.

Tobler, P.N., O'Doherty, J.P., Dolan, R.J., and Schultz, W. (2007). Reward value coding distinct from risk attitude-related uncertainty coding in human reward systems. *Journal of Neurophysiology, 97*(2), 1621–1632.

Tom, S.M., Fox, C.R., Trepel, C., and Poldrack, R.A. (2007). The neural basis of loss aversion in decision-making under risk. *Science, 315*(5811), 515–518.

Tomasello, M., and Vaish, A. (2013). Origins of human cooperation and morality. *Annual Review of Psychology, 64*, 231–255.

Tomasino, D. (2007). The psychophysiological basis of creativity and intuition: Accessing "the zone" of entrepreneurship. *International Journal of Entrepreneurship and Small Business, 4*(5), 528–542.

Trevino, L.K., and Brown, M.E. (2004). Managing to be ethical: Debunking five business ethics myths. *Academy of Management Perspectives, 18*(2), 69–81.

Trivers, R.I. (1971). The evolution of reciprocal altruism. *Quarterly Journal of Biology, 46*(1), 35–57.

Tversky, A., and Kahneman, D. (1974). Judgment under uncertainty: Heuristics and biases. *Science, 185*(4157), 1124–1131.

Tversky, A., and Kahneman, D. (1981). The framing of decisions and the psychology of choice. *Science, 211*(4481), 453–458.

Tversky, A., and Kahneman, D. (1986). Rational choice and the framing of decisions. *Journal of Business, 59*(2), S251–S278.

Van Bavel, J.J., Hall, O.F., and Mende-Siedlecki, P. (2015). The neuroscience of moral cognition: From dual processes to dynamic systems. *Current Opinion in Psychology, 6*(December), 167–172.

Van Bavel, J.J., Packer, D.J., and Cunningham, W.A. (2008). The neural substrates of in-group bias: A functional magnetic resonance imaging investigation. *Psychological Science, 18*(11), 1131–1139.

Van den Bos, K., and Lind, E.A. (2002). Uncertainty management by means of fairness judgments. In M.P. Zanna (ed.), *Advances in Experimental Social Psychology* (Vol. 34, pp. 1–60). San Diego, CA: Academic Press.

Van den Bos, K., and Lind, E.A. (2004). Fairness heuristic theory is an empirical framework: A reply to Arnadottir. *Scandinavian Journal of Psychology, 45*(3), 265–268.

Van Winden, F. (2007). Affect and fairness in economics. *Social Justice Research, 20*(1), 35–52.

Vartanian, O., Mandel, S.R., and Duncan, M. (2011). Money or life: Behavioral and neural context effects on choice under uncertainty. *Journal of Neuroscience, Psychology, and Economics, 4*(1), 25–36.

Veniero, D., Strüber, D., Thut, G., and Herrmann, C.S. (2016). Non-invasive brain stimulation techniques can modulate cognitive processes. *Organizational Research Methods*, 1–32. Doi.org/10.1177/1094428116658960.

Volk, S., and Köhler, T. (2012). Brains and games: Applying neuro-economics to organizational research. *Organization Research Methods*, *15*(4), 522–552.

Von Neumann, J., and Morgenstern, O. (1944). *Theory of Games and Economic Behavior*. Princeton, NJ: Princeton University Press.

Vrba, J., and Robinson, S.E. (2001). Signal processing in magneto-encephalography. *Methods*, *25*(2), 249–271.

Vroom, V.H. (1964). *Work and Motivation*. New York: John Wiley and Sons.

Vugt, M.V., and Ronay, R. (2014). The evolutionary psychology of leadership: Theory, review, and roadmap. *Organizational Psychology Review*, *4*(1), 74–95.

Vul, E., Harris, C., Winkielman, P., and Pashler, H. (2009). Puzzling high correlations in fMRI studies of emotion, personality, and social cognition. *Perspectives on Psychological Science*, *4*(3), 274–290.

Waegeman, A., Declerck, C.H., Boone, C., and Van Hecke, W. (2014). Individual differences in self-control in a time discounting task: An fMRI study. *Journal of Neuroscience, Psychology, and Economics*, *7(2), 65–79.*

Waldman, D.A., and Balthazard, P.A. (2015). Neuroscience of leadership. In D.A. Waldman and P.A. Balthazard (eds), *Organizational Neuroscience (Monographs in Leadership and Management)* (Vol. 7, pp. 189–211). Bingley, UK: Emerald Group Publishing Limited.

Waldman, D.A., Balthazard, P.A., and Peterson, S.J. (2011a). Leadership and neuroscience: Can we revolutionize the way that inspirational leaders are identified and developed? *Academy of Management Perspectives*, *25*(1), 60–74.

Waldman, D.A., Balthazard, P.A., and Peterson, S.J. (2011b). Social cognitive neuroscience and leadership. *The Leadership Quarterly*, *22*(6), 1092–1106.

Waldman, D.A., Wang, D., and Fenters, V. (2016). The added value of neuroscience methods in organizational research. *Organizational Research Methods*. Doi.org/10.1177/1094428116642013.

Waldman, D.A., Wang, D., Hannah, S.T., and Balthazard, P.A. (2017). A neurological and ideological perspective of ethical leadership. *Academy of Management Journal*, *60*(4), 1285–1306.

Waldman, D.A., Wang, D., and Stikic, M. et al. (2015). Neuroscience and team processes. In D.A. Waldman and P.A. Balthazard (eds), *Organizational Neuroscience (Monographs in Leadership and Management*, Vol. 7, pp. 277–294). Bingley, UK: Emerald Group Publishing Limited.

Waldman, D.A., Ward, M.K., and Becker, W.J. (2017). Neuroscience in organizational behavior. *Annual Review of Organizational Psychology and Organizational Behavior*, *4*, 425–44.

Wallas, G. (1926). *The Art of Thought*. London: Jonathan Cape.

Walsh, V. and Conwey, A. (2000). Transcranial magnetic stimulation and cognitive neuroscience. *Nature Reviews Neuroscience*, *1*(1), 73–79.

Walter, H (2012). Social cognitive neuroscience of empathy: Concepts, circuits, and genes. *Emotion Review*, *4*(1), 9–17.

Walter, H., Abler, B., Ciaramidaro, A., and Erk, S. (2005). Motivating forces of human actions: Neuroimaging reward and social interaction. *Brain Research*, *67*(5), 368–381.

Ward, M.K., and Becker, W.J. (2013). Organizational neuroscience. *The Industrial Organizational Psychologist*, *51*, 94–97.

Ward, M.K., Volk, S., and Becker, W.J. (2015). An overview of organizational neuroscience. In D.A. Waldman and P.A. Balthazard (eds), *Organizational Neuroscience: Monographs in Leadership and Management* (Vol. 7, pp. 17–50). Bingley, UK: Emerald Group Publishing Limited.

Ward, T.B. (2007). Creative cognition as a window on creativity. *Methods*, *42*(1), 28–37.

Wardle, M.C., Fitzgerald, D.A., and Angstadt, M. et al. (2013). The caudate signals bad reputation during trust decisions. *PLOS One*, *8*(6), 1–9.

Watson, K., and Platt, M.L. (2006). Fairness and the neurobiology of social cognition: Commentary on nonhuman species' reactions to inequity and their implications for fairness by Sarah Brosnan. *Social Justice Research*, *19*(2), 186–193.

Weber, E.U., and Johnson, E.J. (2009). Mindful judgment and decision making. *Annual Review of Psychology*, *60*, 53–85.

Weiss, H.M., and Cropanzano, R. (1996). Affective events theory: A theoretical discussion of the structure, causes and consequences of affective experiences at work. In B.M. Staw and L.L. Cummings (eds), *Research in Organizational Behavior* (Vol. 18, pp. 1–74), Greenwich, CT: Elsevier Science/JAI Press.

Welling, H. (2007). Four mental operations in creative cognition: The importance of abstraction. *Creativity Research Journal*, *19*(2/3), 163–177.

White, R.W. (1959). Motivation reconsidered: The concept of competence. *Psychological Review*, *66*(5), 297–333.

Wicker, B., Keysers, C., and Plailly, J. et al. (2003). Both of us disgusted in my insula: The common neural basis of seeing and feeling disgust. *Neuron*, *40*(3), 655–664.

Williamson, O. (1993). Calculativeness, trust, and economic organization. *Journal of Law and Economics, 36*(1), 453–486.

Winston, J.S., Strange, B.A., Doherty, J., and Dolan, R.J. (2002). Automatic and intentional brain responses during evaluation of trustworthiness of faces. *Nature Neuroscience, 5*(3), 277–283.

Wood, R.C., Levine, D.S., Cory, G.A., and Wilson, D.R. (2015). Evolutionary neuroscience and motivation in organizations. In D.A. Waldman and P.A. Balthazard (eds), *Organizational Neuroscience: Monographs in Leadership and Management* (Vol. 7, pp. 143–167). Bingley, UK: Emerald Group Publishing Limited.

Wout, V.M., Kahn, R.S., Sanfey, A.G., and Aleman, A. (2006). Affective state and decision-making in the ultimatum game. *Experimental Brain Research, 169*(4), 564–568.

Wright, N.D., Symmonds, M., Fleming, S.M., and Dolan, R.J. (2011). Neural segregation of objective and contextual aspects of fairness. *Journal of Neuroscience, 31*(14), 5244–5252.

Wu, X., Yang, W., and Tong, D. et al. (2015). A meta-analysis of neuroimaging studies on divergent thinking using activation likelihood estimation. *Human Brain Mapping, 36*(7), 2703–2718.

Wu, Y., Leliveld, M.C., and Zhou, X. (2011). Social distance modulates recipients' fairness considerations in the dictator game: An ERP study. *Biological Psychology, 88*(2/3), 253–262.

Wu, Y., Zhou, Y., and Van Dijk, E.V. et al. (2011). Social comparison affects brain responses to fairness in asset division: An ERP study with the ultimatum game. *Frontiers in Human Neuroscience, 5*, 131–141.

Yang, J., Li, H., and Zhang, Y. et al. (2007). The neural basis of risky decision-making in a blackjack task. *NeuroReport, 18*(14), 1507–1510.

Yaniv, D. (2011). Revisiting Morenian psychodramatic *encounter* in light of contemporary neuroscience: Relationship between empathy and creativity. *The Arts in Psychotherapy, 38*(1), 52–58.

Yoruk, S., and Runco, M.A. (2014). The neuroscience of divergent thinking. *Activitas Nervosa Superior, 56*(1/2), 1–16.

Young, L., and Dungan, J. (2012). Where in the brain is morality? Everywhere and maybe nowhere. *Journal of Social Neuroscience, 7*(1), 1–10.

Young, L., and Saxe, R. (2008). An fMRI investigation of spontaneous mental state inference of moral judgment. *Journal of Cognitive Neuroscience, 21*(7), 1396–1405.

Zaccaro, S.J. (2007). Trait-based perspective of leadership. *American Psychologist, 62*(1), 6–16.

Zahra, S.A., Filatotchev, I., and Wright, M. (2009). How do threshold firms captain corporate entrepreneurship? The role of boards and absorptive capacity. *Journal of Business Venturing, 24*(3), 248–260.

Zak, P.J. (2004). Neuroeconomics. *Philosophical Transactions of the Royal Society, 359*(1451), 1737–1748.

Zak, P.J., Kurzban, R., and Matzner, W.T. (2005). Oxytocin is associated with human trustworthiness. *Hormones and Behavior, 48*(5), 522–527.

Zaki, J., and Mitchell, J.P. (2011). Equitable decision making is associated with neural markers of intrinsic value. *Proceedings of the National Academy of Sciences, 108*(49), 19761–19766.

Zaki, J., and Ochsner, K.N. (2012). The neuroscience of empathy: Progress, pitfalls and promise. *Nature Neuroscience, 15*(5), 675–80.

Zhou, J., and Hoever, I.J. (2014). Research on workplace creativity: A review and redirection. *Annual Review of Organizational Psychology and Organizational Behavior, 1*, 333–359.

Zink, C.F., Tong, Y., and Chen, Q. et al. (2008). Know your place: Neural processing of social hierarchy in humans. *Neuron, 58*(2), 273–283.

Subject index

Author index